LONDON MATHEMATICAL SOCIETY LECTURE NOTE SERIES

Editor: PROFESSOR G. C. SHEPHARD, University of East Anglia

This series publishes the records of lectures and seminars on advanced topics in mathematics held at universities throughout the world. For the most part, these are at postgraduate level either presenting new material or describing older material in a new way. Exceptionally, topics at the undergraduate level may be published if the treatment is sufficiently original.

Prospective authors should contact the editor in the first instance.

Already published in this series

1. General cohomology theory and K-theory, PETER HILTON.
2. Numerical ranges of operators on normed spaces and of elements of normed algebras, F. F. BONSALL and J. DUNCAN.
3. Convex polytopes and the upper bound conjecture, P. McMULLEN and G. C. SHEPHARD.
4. Algebraic topology: A student's guide, J. F. ADAMS.
5. Commutative algebra, J. T. KNIGHT.
6. Finite groups of automorphisms, NORMAN BIGGS.
7. Introduction to combinatory logic, J. R. HINDLEY, B. LERCHER and J. P. SELDIN.
8. Integration and harmonic analysis on compact groups, R. E. EDWARDS.
9. Elliptic functions and elliptic curves, PATRICK DU VAL.
10. Numerical ranges II, F. F. BONSALL and J. DUNCAN.
11. New developments in topology, G. SEGAL (ed.).
12. Symposium on complex analysis Canterbury, 1973, J. CLUNIE and W. K. HAYMAN (eds.).
13. Combinatorics, Proceedings of the British combinatorial conference 1973, T. P. McDONOUGH and V. C. MAVRON (eds.).
14. Analytic theory of abelian varieties, H. P. F. SWINNERTON-DYER.
15. An introduction to topological groups, P. J. HIGGINS.
16. Topics in finite groups, TERENCE M. GAGEN.
17. Differentiable germs and catastrophes, THEODOR BRÖCKER and L. LANDER.
18. A geometric approach to homology theory, S. BUONCRISTIANO, C. P. ROURKE and B. J. SANDERSON.
19. Graph theory, coding theory and block designs, P. J. CAMERON and J. H. VAN LINT.
20. Sheaf theory, B. R. TENNISON.
21. Automatic continuity of linear operators, ALLAN M. SINCLAIR.
22. Presentations of groups, D. L. JOHNSON.
23. Parallelisms of complete designs, PETER J. CAMERON.
24. The topology of Stiefel manifolds, I. M. JAMES.
25. Lie groups and compact groups, J. F. PRICE.
26. Transformation groups: Proceedings of the conference in the University of Newcastle upon Tyne, August 1976, CZES KOSNIOWSKI.
27. Skew field constructions, P. M. COH

London Mathematical Society Lecture Note Series. 28

Brownian motion, Hardy spaces and bounded mean oscillation

K. E. Petersen

Associate Professor of Mathematics

University of North Carolina at Chapel Hill

CAMBRIDGE UNIVERSITY PRESS

CAMBRIDGE

LONDON NEW YORK MELBOURNE

Published by the Syndics of the Cambridge University Press
The Pitt Building, Trumpington Street, Cambridge CB2 1RP
Bentley House, 200 Euston Road, London NW1 2DB
32 East 57th Street, New York, N.Y. 10022, USA
296 Beaconsfield Parade, Middle Park, Melbourne 3206, Australia

Library of Congress Cataloging in Publication Data

Petersen, Karl Endel, 1943-

Brownian motion, Hardy spaces, and bounded mean oscillation.

(London Mathematical Society lecture note series; 28)

Bibliography: p.
1. Brownian motion processes. 2. Hardy spaces.
3. Oscillations. I. Title. II. Series: London Mathematical Society.
Lecture note series; 28.
QA274.75.P47 519.2'8 76-46860

ISBN: 0 521 21512 9

First published 1977

Printed in Great Britain
at the University Press, Cambridge

For F. A. P. and E. A. P.

Limits we did not set condition all we do.

Matthew Arnold

Contents

Preface

These notes address the connection between two subjects, and they are thus intended to form an introduction to both but to be about neither. The discoveries of Fefferman and Stein about H^p and BMO have interacted fruitfully with a great deal of work on the analogous ideas in martingale theory; the main goal of the following pages is an explanation of the fundamental result of Burkholder, Gundy, and Silverstein, which forms the bridge between these two areas of investigation. The exposition is at as elementary a level as possible, and it is intended in particular to be accessible to graduate students with a basic knowledge of measure theory, complex analysis and functional analysis. For the sake of those not familiar with probability theory, many probabilistic results are introduced and proved as needed, and there is a chapter without proofs on Brownian motion. Again, for those not on everyday terms with classical function theory, a survey of results on the maximal, square, and Littlewood-Paley functions is included, and function-theoretic arguments are given and estimates made in considerable detail. The discussion is restricted mainly to the case of the unit disk in the complex plane. I hope that one who reads these notes will find that Garsia's book, the papers of Fefferman and Stein, and the writings of Burkholder, Davis, Gundy, Herz, Silverstein, et al. on these topics are easily approachable.

I originally organized this material for a series of seminar talks that I gave at U. N. C. in the fall of 1975, and it is a pleasure to thank the participants for all that they taught me. The review of the basic properties of Brownian motion is based in part on lectures given by S. Kakutani, and I am grateful to R. F. Gundy for an outline of much of the material in Chapter 7. I would also like to thank Janet Farrell for her quick and accurate typing of the original manuscript and the National Science Foundation for support during the period that these notes were in preparation.

Chapel Hill, N. C.
March 1976

Karl Petersen

1·Introduction

In 1915 G. H. Hardy, answering a question of Bohr and Landau, investigated properties of the mean over a circle of the modulus $|F|$ of an analytic function F which were similar to those of the maximum value of $|F|$ over a disk. He found that his results applied also to $|F|^p$ for $p > 0$, and thus was founded the theory of H^p spaces. Since then these Hardy spaces have been the object of much research, and their connections with such diverse subjects as classical function theory (especially the boundary behavior of analytic functions), potential theory (including the theory of harmonic functions and partial differential equations), Fourier series, functional analysis, and operator theory (for example Beurling's work on invariant subspaces of the shift operator) have been developed in considerable detail.

An entirely new line of investigation for the Hardy spaces was uncovered in 1971 by Burkholder, Gundy, and Silverstein when they showed that for $0 < p < \infty$ an analytic function $F = u + i\tilde{u}$ is in H^p if and only if the maximal function of u is in L^p. Surprisingly, their arguments were probabilistic in nature, being carried out by manipulation of Brownian motion in the complex plane. Their result showed that the Hardy spaces could be characterized in real-variable terms and thus H^p theory could be easily extended to higher dimensions and more general kinds of spaces.

Such a generalization and extension were accomplished in 1972 by Fefferman and Stein with analytic rather than probabilistic arguments, in which a careful study of the Lusin S-function (which is also called the 'square function' and 'area function') played an important part. (Their work built on the efforts of many authors in addition to Burkholder, Gundy, and Silverstein, including, for example, Calderón, Segovia, Stein and Weiss, and Zygmund; it would take us too far afield, however, to attempt to indicate in a historically correct manner the contributions of the many researchers who have dealt with these problems.) They were able to extend the Burkholder-Gundy-Silverstein result to higher dimensions, give a real-variable characterization of H^p, and identify the dual space of H^1 as the space BMO of all functions of bounded mean oscillation. In particular, they showed that for any analytic function $F = u + i\tilde{u}$ and any p with $0 < p < \infty$, the following seven statements are equivalent: (1) $F \in H^p$ (for $p \geq 1$ this is equivalent to the existence of an L^p boundary function); (2) the classical maximal function $N_\sigma F$ (whose values are the least upper bounds of $|F|$ over Stolz domains) is in L^p; (3) the Lusin S-function $S(F)$ of F is in L^p; (4) $S(u) \in L^p$ (trivial, since $S(u) = S(F)$); (5) $N_\sigma u \in L^p$; (6) (for $p \geq 1$) if ϕ is the boundary function of u and $\mathcal{P}_r$ is the Poisson kernel, then $\sup_{0\leq r<1} |\mathcal{P}_r * \phi(x)| \in L^p$; and (7) (for $p \geq 1$) $\sup_{0\leq r<1} |\psi_r * \phi(x)| \in L^p$ for each reasonably smooth approximate identity $\{\psi_r\}$.

The duality between H^1 and BMO makes possible other real-variable characterizations of H^p. Denote by Γ the unit circle in the complex plane and by m normalized Lebesgue measure on Γ. A real-valued function $a \in L^\infty(\Gamma)$ is called an <u>atom</u> in case

(i) a is supported on an interval I and $\|a\|_\infty \leq \frac{1}{m(I)}$,

(ii) a takes no more than two non-zero values, and

(iii) $\int a\,dm = 0$.

Let u be a real-valued function in $L^1(\Gamma)$. Then u is the real part of the boundary function of some $f \in H^1$ if and only if there are a sequence $a_1, a_2, \ldots$ of atoms and a sequence $\lambda_1, \lambda_2, \ldots$ of real numbers such that $\sum |\lambda_n| < \infty$ and $u = \int u\,dm + \sum \lambda_n a_n$. Moreover, letting $\lambda(u)$ equal the infimum of $\sum |\lambda_n|$ over all such sequences $\lambda_1, \lambda_2, \ldots$, there are universal constants c_1 and c_2 such that $c_1 \|f\|_1 \leq |\int u\,dm| + \lambda(u) \leq c_2 \|f\|_1$. This striking result is also due to Fefferman. A very simple proof has been given by Axler [2], and Coifman [10] has published a proof for the case $0 < p \leq 1$.

Probabilistic proofs are possible in this area of analysis because of the connection between probability theory and potential theory, which is seen in Kakutani's theorem equating harmonic measure with the hitting probabilities of Brownian motion. For a point $z \in D$ in the unit disk D

of the complex plane and a measurable subset $A \subset \Gamma$ of the boundary Γ of D, the harmonic measure $\omega_z(A)$ of A at z is defined to be the value at z of the harmonic function in D whose boundary values are 1 on A and 0 on $\Gamma \backslash A$; thus

$$\omega_z(A) = \frac{1}{2\pi} \int_A \mathcal{P}(r, \theta - t)dt,$$

where $z = re^{i\theta}$ and $\mathcal{P}(r, \theta - t) = \dfrac{1 - r^2}{1 - 2r\cos(\theta - t) + r^2}$ is the Poisson kernel. Kakutani proved in 1944 that $\omega_z(A)$ equals the probability that a Brownian traveler starting from the point z, at the time that he first hits Γ, hits Γ in a point of A. Even more, if S is any bounded, connected, open subset of R^n with smooth boundary ∂S, f is continuous on ∂S, $\gamma_{z,t}$ is a Brownian motion starting at $z \in S$, and $T(z) = \inf\{t \geq 0 : \gamma_{z,t} \notin S\}$, then $u(z) = E(f(\gamma_{z,T(z)}))$ is harmonic in S and $\lim_{z \to x} u(z) = f(x)$ for each $x \in S$ (that is, u is the solution in S of the Dirichlet problem with boundary values f). Besides its applications to analysis, this connection also led to the development of probabilistic potential theory by Doob, Hunt, and others.

Similarly, many of the classical arguments involving the maximal and square functions have been carried forward also in a probabilistic setting, and now even the inequalities involving H^p and BMO norms are seen to apply also to abstract martingales.

The main purpose of these notes is an investigation of the relationship of these new martingale inequalities to the analytic results which they mirror, and therefore the major portion is taken up by an exposition of the theorem of Burkholder, Gundy, and Silverstein. It will be seen that in many of the proofs translations are made from the probabilistic to the analytic setting by changes of variables in accordance with Kakutani's theorem. We begin with a historical survey of results related to the Hardy-Littlewood maximal function, the Lusin S-function, and the Littlewood-Paley g-function, and Chapter 3 gives a short list of relevant definitions and theorems related to Brownian motion. The concluding chapter justifies the probabilistic definitions of H^p and BMO and includes a proof of the H^1-BMO duality theorem via continuous-time martingales. I hope that the reader will go then to the original literature for the full details of this developing story.

2 · The maximal, square and Littlewood-Paley functions

D denotes the unit disk $\{z \in \mathbf{C} : |z| < 1\}$ in the complex plane, $\Gamma = \{e^{i\theta} : 0 \leq \theta < 2\pi\}$ its boundary, and m normalized Lebesgue measure on Γ. For a function f defined on D, $0 < p < \infty$, and $0 \leq r < 1$, we define

$$M_p(r, f) = [\int_0^{2\pi} |f(re^{i\theta})|^p dm(\theta)]^{1/p} .$$

An analytic function f on D is said to be of the class H^p in case

$$\sup_{r<1} M_p(r, f) = \|f\|_p < \infty .$$

The space of bounded analytic functions on D, with the supremum norm, is denoted by H^∞. Each (real-valued) harmonic function u on D has a unique conjugate function $\tilde{u}$ determined by the conditions that $\tilde{u}(0) = 0$ and $u + i\tilde{u}$ is analytic on D.

For a fixed σ with $0 < \sigma < 1$, each point $e^{i\theta} \in \Gamma$ determines a Stolz domain $\Omega_\sigma(\theta)$, as shown in the figure:

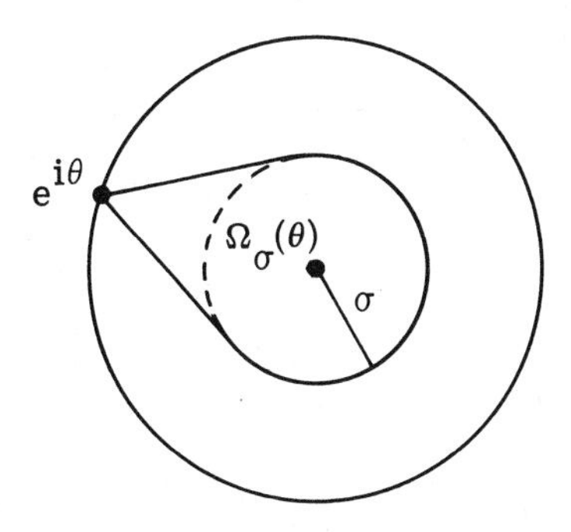

$\Omega_\sigma(\theta)$ is the interior of the convex hull of the circle of radius σ with center at the origin and the point $e^{i\theta}$.

Given any function f defined on D, the (classical) maximal function $N_\sigma f$ of f is defined by

$$N_\sigma f(e^{i\theta}) = \sup \{ |f(z)| : z \in \Omega_\sigma(\theta) \}.$$

The <u>square function</u> $S_\sigma F$ of an analytic function F on D is

$$S_\sigma F(e^{i\theta}) = [\int\int_{\Omega_\sigma(\theta)} |F'(z)|^2 dxdy]^{\frac{1}{2}} .$$

The square of $S_\sigma F(e^{i\theta})$ equals the area of the image of $\Omega_\sigma(\theta)$ under F. The <u>Littlewood-Paley function</u> g_F of an analytic function F on D is defined by

$$g_F(e^{i\theta}) = [\int_0^1 (1 - r)|F'(re^{i\theta})|^2 dr]^{\frac{1}{2}} .$$

Each of these functions measures in its own way the growth of the given function near the boundary of D. It has been proved that for an analytic function F, and for almost all $e^{i\theta} \in \Gamma$, the existence of a nontangential limit for F at $e^{i\theta}$ is equivalent to the finiteness of $N_\sigma F(e^{i\theta})$, or alternatively, to that of $S_\sigma F(e^{i\theta})$. Also, many inequalities among the norms of F, $N_\sigma F$, $S_\sigma F$, and g_F have been established, and these results have been carried over to the theory of martingales. Some idea of the history of the subject may be obtained from the following list of several of the important discoveries concerning these functions. The obvious conventions with regard to inequalities are in force when one side happens to be infinite.

2.1 G. H. Hardy and J. E. Littlewood, 1930 [24]

For each p with $0 < p < \infty$ there is a constant c_p such that whenever $F = u + i\tilde{u}$ is analytic on D,

$$\int_0^{2\pi} |N_\sigma F(e^{i\theta})|^p dm(\theta) \le c_p \sup_{0 \le r < 1} \int_0^{2\pi} |F(re^{i\theta})|^p dm(\theta) ,$$

or in more compact notation, $\|N_\sigma F\|_{L^p(\Gamma)} \le c_p \|F\|_{H^p}$. Hence also $\|N_\sigma u\|_{L^p(\Gamma)} \le c_p \|F\|_{H^p}$, and so this result comprises half of the Burkholder-Gundy-Silverstein Theorem, whose probabilistic proof we will examine below. (The Hardy-Littlewood proof relied on a combinatorial lemma and imagery referring to the game of cricket.)

In the same paper Hardy and Littlewood also considered a maximal function $M\phi$ defined for nonnegative $\phi \in L^1(\Gamma)$ by

$$M\phi(e^{i\theta}) = \sup_{e^{i\theta} \in I} \frac{1}{m(I)} \int_I \phi(e^{i\theta})dm(\theta)$$

and proved that for each $p > 1$ there is a constant c_p such that

$$\|M\phi\|_p \leq c_p \|\phi\|_p .$$

2.2 **N. Lusin,** 1930 [41]

For each σ with $0 < \sigma < 1$ there is a constant A_σ such that if $F(z) = \sum_{n=0}^{\infty} a_n z^n$ is in H^2, then

$$\int_0^{2\pi} [S_\sigma F(e^{i\theta})]^2 dm(\theta) \leq A_\sigma \sum_{n=0}^{\infty} |a_n|^2 .$$

It follows, then, that $S_\sigma F(e^{i\theta}) < \infty$ for almost all θ.

2.3 **J. Marcinkiewicz and A. Zygmund,** 1938 [42]

(i) For each σ with $0 < \sigma < 1$ and p with $0 < p < \infty$ there is a constant $A_{p,\sigma}$ such that

$$\|S_\sigma F\|_{L^p(\Gamma)} \leq A_{p,\sigma} \|F\|_{H^p}$$

for each analytic function F on D.

(ii) If F is analytic on D and has a nontangential limit at every point of a subset E of Γ which has positive measure, then $S_\sigma F(e^{i\theta}) < \infty$ a.e. on E.

D. Spencer, 1943 [52]

Let F be analytic on D. If $E \subset \Gamma$ has positive measure and for each $e^{i\theta} \in E$ there is $\sigma = \sigma(\theta)$ with $0 < \sigma < 1$ such that $S_\sigma F(e^{i\theta}) < \infty$, then F has a nontangential limit at almost every point of E.

2.4 **J. E. Littlewood and R. E. A. C. Paley, 1931, 1936, 1937 [37, 38, 39]**

For each p with $0 < p \leq \infty$ there is a constant C_p such that if F is analytic on D then

$$\|g_F\|_{L^p(\Gamma)} \leq C_p \|F\|_{H^p}.$$

Conversely, for $1 < p < \infty$ there is a constant B_p such that for each analytic F on D

$$\|F\|_{H^p} \leq B_p \|g_F\|_{L^p(\Gamma)}.$$

J. Marcinkiewicz and A. Zygmund, 1938 [42]

For each σ with $0 < \sigma < 1$ there is a constant A_σ such that if F is analytic on D then

$$g_F(e^{i\theta}) \leq A_\sigma S_\sigma F(e^{i\theta}) \quad \text{for all } \theta \in [0, 2\pi).$$

G. Gasper, Jr., 1968 [22]

extended some of these results on g_F and $S_\sigma F$ to higher dimensions.

2.5 **M. Riesz, 1920 [49]**

For $0 < p < \infty$ let h^p denote the space of all real-valued harmonic functions u in D for which

$$\|u\|_{h^p} = \sup_{0 \leq r < 1} M_p(r, u) < \infty,$$

and let h^∞ denote the space of all bounded real-valued harmonic functions on D with the supremum norm.

Then for each p with $1 < p < \infty$ there is a constant C_p such that

$$\|\tilde{u}\|_{h^p} \leq C_p \|u\|_{h^p} \quad \text{for all } u \in h^p.$$

A. N. Kolmogorov, 1925 [33]

For each p with $0 < p < 1$ there is a constant C_p such that whenever $u \in h^1$,

$$\|\tilde{u}\|_{h^p} \leq C_p \|u\|_{h^1}.$$

B. Davis, 1974 [12]

used Brownian motion to give a proof of Kolmogorov's result and at the same time determined the best possible values of the constants C_p.

The M. Riesz inequalities are proved below (5.2) as a corollary of the proof of the Burkholder-Gundy-Silverstein Theorem.

2.6 The dual space of H^p

The above result of M. Riesz implies that if $g \in L^q(\Gamma)$ for some $q > 1$, so that there is a Fourier expansion $g(e^{i\theta}) \sim \sum_{n=-\infty}^{\infty} a_n e^{in\theta}$, then the 'analytic projection' $\sum_{n=0}^{\infty} a_n e^{in\theta}$ of g is in H^q. This fact was used in the identification of the dual space of H^p for $1 \leq p < \infty$.

Denote by H_0^q the set of all those $F \in H^q$ for which $F(0) = 0$. Then for $1 \leq p < \infty$, $(H^p)^*$ is isometrically isomorphic to the quotient Banach space L^q/H_0^q, where q is the conjugate index to p determined by the equation $\frac{1}{p} + \frac{1}{q} = 1$. The action of each linear functional $\phi \in (H^p)^*$ is given by

$$\langle f, \phi \rangle = \int_0^{2\pi} f(e^{i\theta})\overline{g(e^{i\theta})}\, dm(\theta) \quad (f \in H^p)$$

for some $g \in H^q$, which is unique in case $1 < p < \infty$. Also, for each p with $1 < p < \infty$ there is a constant C_p such that

$$\|\phi\| \leq \|g\|_q \leq C_p \|\phi\|,$$

where ϕ and g are related as above. Therefore $(H^p)^*$ is 'essentially' H^q for $1 \leq p < \infty$.

Most of these results are due to

A. E. Taylor, 1950, 1951 [53, 54, 55].

2.7 The martingale approach

Many of these ideas can be formulated in the context of abstract martingale theory. In fact, results that are proved in that setting frequently carry over to the classical situation when a Brownian traveler is employed to sample the values of an analytic function.

If $f_1, f_2, \ldots$ is a martingale (so that $E(f_{n+1} | f_1, \ldots, f_n) = f_n$; further details are in 3.4), the maximal and square functions of $\{f_n\}$ are defined by

$$f^* = \sup_n |f_n| \quad \text{and}$$

$$Sf = [\sum_{k=1}^{\infty} (f_{k+1} - f_k)^2]^{\frac{1}{2}},$$

respectively.

D. L. Burkholder, 1966 [4]

For each p with $1 < p < \infty$ there are constants c_p and C_p such that

$$c_p \|Sf\|_p \leq \|f^*\|_p \leq C_p \|Sf\|_p .$$

B. Davis, 1970 [11]

extended this result to the case $p = 1$.

D. L. Burkholder and R. F. Gundy, 1970 [6]

extended this result to the case $0 < p < 1$ in case the martingale satisfies a certain regularity condition.

They and Silverstein [7] also considered a continuous-time square function operator $\mathcal{S}$ - if $\gamma_{0,t}$ is Brownian motion in D starting at 0, if τ is the first hitting time of $\gamma_{0,t}$ on Γ, and if u is harmonic on D, then

$$\mathcal{S}u(\omega) = [\int_0^{\tau(\omega)} |\nabla u(\gamma_{0,t}(\omega))|^2 dt]^{\frac{1}{2}}$$

- and proved that for each p with $0 < p < \infty$ there are constants c_p and C_p such that

$$c_p \| \mathcal{S}u \|_p \leq \|u^*\|_p \leq C_p \| \mathcal{S}u \|_p$$

(where $u^*(\omega) = \sup_{0 \leq t < \tau(\omega)} |u(\gamma_{0,t}(\omega))|$). This implies that if u is harmonic on D and $u(0) = \tilde{u}(0) = 0$, then for $0 < p < \infty$

$$\|\tilde{u}^*\|_p \leq C_p \|u^*\|_p.$$

2.8 D. L. Burkholder, R. F. Gundy, and M. L. Silverstein, 1971 [7]

For each σ with $0 < \sigma < 1$ and each p with $0 < p < \infty$ there are constants $c_{\sigma,p}$ and $C_{\sigma,p}$ such that whenever u is harmonic on D and $F = u + i\tilde{u}$ is its analytic completion,

$$c_{\sigma,p} \|N_\sigma u\|_{L^p(\Gamma)} \leq \|F\|_{H^p} \leq C_{\sigma,p} \|N_\sigma u\|_{L^p(\Gamma)}.$$

Note that the left-hand inequality is the result of Hardy and Littlewood mentioned earlier (2.1).

We may restate this theorem as follows: For $0 < p < \infty$, $F \in H^p$ if and only if the maximal function of $\mathrm{Re}(F)$ is in L^p. Therefore one can tell whether a given function in $L^p(\Gamma)$ is the real part of an H^p function by checking whether or not its maximal function is also in L^p - there is no need first to compute the conjugate function.

Since the question answered by this result is of such importance, some further background may be called for. For $0 < p < \infty$, each H^p function F determines via nontangential limits a boundary function which is in $L^p(\Gamma)$. Moreover, for $p \geq 1$ each H^p function is the Poisson integral of its boundary values; and if $\phi \in L^p(\Gamma)$ is given and the Poisson integral F of ϕ is analytic in D, then $F \in H^p$. However, for $p \leq 1$ certain crucial difficulties appear, the central one being that the harmonic conjugate of an h^1 function need not belong to h^1. For a long time it was not known how to determine, given $\phi \in L^1(\Gamma)$, whether or not $\int \mathcal{P}(r, \theta - t)\phi(t)dm(t)$ is in H^1 (by looking at ϕ alone). In particular, for real-valued $\phi \in L^1(\Gamma)$, $\int \mathcal{P}(r, \theta - t)\phi(t)dm(t)$ always defines a harmonic function u on D, but it was hard to characterize those functions ϕ for which $\tilde{u} \in h^1$. The Burkholder-Gundy-Silverstein and Fefferman-Stein theorems have settled these problems by specifying exactly when $u \in \mathrm{Re}(H^1)$.

2.9 **C. Fefferman and E. M. Stein, 1972 [19]**

Let $F = u + i\tilde{u}$ be analytic on D and $0 < p < \infty$. Then the following are equivalent:

(i) $F \in H^p$,

(ii) $N_\sigma F \in L^p$,

(iii) $S_\sigma F \in L^p$,

(iv) $S_\sigma u \in L^p$,

(v) $N_\sigma u \in L^p$,

(vi) (for $p \geq 1$) $\sup_{0 \leq r < 1} |\mathcal{P}_r * \phi(e^{i\theta})| \in L^p$ (where ϕ is the boundary function of u and $\mathcal{P}_r$ is the Poisson kernel), and

(vii) (for $p \geq 1$) $\sup_{0 \leq r < 1} |\psi_r * \phi(e^{i\theta})| \in L^p$ for any reasonably smooth approximate identity $\{\psi_r\}$.

2.10 **L. Carleson, 1958, 1962 [8, 9]**

A measure μ on D is called a <u>Carleson measure</u> in case there is a constant A such that

$$\mu\{re^{i\theta} : 1 - h \leq r < 1,\ \theta_0 \leq \theta < \theta_0 + h\} \leq Ah$$

for each h with $0 < h < 1$ and each θ_0 with $0 \leq \theta_0 < 2\pi$.

Let μ be a finite measure on D and $0 < p < \infty$. Then μ is a Carleson measure if and only if there is a constant C such that

$$\left| \int_D |F(z)|^p d\mu(z) \right|^{1/p} \leq C \|F\|_{H^p} \quad \text{for all } F \in H^p.$$

This result was one step in the proof (1961) of the <u>Corona Conjecture</u> (Kakutani, 1944): The point evaluations are dense in the maximal ideal space of H^∞.

2.11 **C. Fefferman and E. M. Stein, 1972 [19]**

For a real-valued function $\phi \in L^2(\Gamma)$ with $\int \phi\, dm = 0$ the following statements are equivalent:

(i) There is a constant C such that

$$\left| \int_0^{2\pi} F(e^{i\theta})\phi(e^{i\theta})dm(\theta) \right| \le C \|F\|_{H^1} \quad \text{for all } F \in H^2$$

(and so this action of ϕ extends by continuity to a continuous linear functional on H^1);

(ii) $\phi = \psi_1 + \tilde{\psi}_2$ for some real-valued $\psi_1, \psi_2 \in L^\infty(\Gamma)$;

(iii) ϕ has <u>bounded mean oscillation:</u>

$$\|\phi\|_{BMO} = \sup_I \frac{1}{m(I)} \int_I \left|\phi - \int_I |\phi|\right| < \infty;$$

(iv) If u is the Poisson integral of ϕ, then $(1 - |z|^2)|\nabla u|^2 dxdy$ is a Carleson measure on D.

Moreover, every continuous linear functional on H^1 is determined by a unique $\phi \in BMO$ according to (i), and the BMO and $(H^1)^*$ norms are equivalent.

2.12 Probability

The study of the martingale versions of these ideas has been carried forward by Burkholder, Davis, Garsia, Getoor, Gundy, Herz, Sharpe, Silverstein, et al. The techniques associated with these investigations are leading to the discovery of some interesting probabilistic proofs of other theorems in analysis. For instance, Davis [13] has used Brownian motion to prove Picard's little theorem (to the effect that a non-constant entire function takes every value with at most one exception).

3·Brownian motion

The purpose of this chapter is to define and construct Brownian motion and to list enough of its properties so that we will be able to carry out the subsequent arguments which rely on them. We omit most proofs and suggest that the interested reader refer to [14, 18, 20, 27, 34, 36, 40, 43, 44, or 58] for more details.

The strange spontaneous movements of small particles suspended in liquid were first studied by Robert Brown, an English botanist, in 1828, although they had apparently been noticed much earlier by other scientists, including even Leeuwenhoek. L. Bachelier gave the first mathematical description of the phenomenon in 1900, going so far as to note the Markov property of the process. In 1905 A. Einstein and, independently and around the same time, M. v. Smoluchowski developed physical theories of Brownian motion which could be used, for example, to determine molecular diameters. It is interesting that Einstein says that at that time he had never heard of the actual observations of Brownian motion, and that he happened to deduce the existence of such a process in the course of some purely theoretical work on statistical mechanics and thermodynamics [50, p. 47]. The mathematical theory of Brownian motion was invented in 1923 by N. Wiener, and accordingly the Brownian motion that we will be working with is frequently called the Wiener process. This theory makes no pretense of having any real connection with the physical Brownian motion - no particle could follow a typical sample path of the Wiener process; instead, the applications of the Wiener process are to the general theory of stochastic processes and such diverse fields as information theory, harmonic analysis, partial differential equations, and, as we shall see, complex analysis. G. E. Uhlenbeck and L. S. Ornstein in 1930 [56] constructed a process somewhat similar to the Wiener process but more closely related to physical reality, for example in that the momentum of each particle is taken into account.

3.1 Definition

A <u>one-dimensional Brownian motion process</u> is a family $\{x_t : t \geq 0\}$ of real-valued measurable functions x_t on a probability space $(\Omega, \mathcal{B}, P)$ with the following properties:

(i) $x_0(\omega) = 0$ a. e.

(ii) for each $s, t \geq 0$, $x_{s+t} - x_t$ is normally distributed with mean 0 and variance s; this means that for $a \leq b$,

$$P\{\omega \in \Omega : a < x_{s+t}(\omega) - x_t(\omega) < b\} = \frac{1}{\sqrt{(2\pi s)}} \int_a^b e^{-u^2/2s} du.$$

(iii) If $0 \leq t_0 < t_1 < \ldots < t_n$, then the random variables $x_{t_{i+1}} - x_{t_i}$, for $i = 0, 1, \ldots, n-1$, are independent. (Recall that random variables $f_1, \ldots, f_n$ are said to be independent if the σ-algebras that they generate are independent. Thus if $\mathcal{A}_k$ is the smallest σ-algebra with respect to which f_k is measurable, then $\mathcal{A}_1, \ldots, \mathcal{A}_n$ are independent in the following sense: if $A_k \in \mathcal{A}_k$ for $k = 1, \ldots, n$, then $P(A_1 \cap A_2 \cap \ldots \cap A_n) = P(A_1)P(A_2) \ldots P(A_n)$.)

The number $x_t(\omega)$ is thought of as representing the position on the real line at time t of a Brownian traveler ω who at time 0 is located at the origin. The randomness of the motion is invested in the choice of ω from Ω according to the probability measure P. A point $\omega \in \Omega$ may be thought of as the entire history of a given Brownian traveler or, equally as well, as the traveler himself.

Motivation for condition (ii) may be found in the observation that $x_{s+t} - x_t$, which represents the displacement of a Brownian traveler over a time interval of length s, may be supposed to consist of the sum of many small, random, independent displacements, and so, by the Central Limit Theorem, should conform to a Gaussian distribution.

It can be proved that if $\{x_t\}$ is a Brownian motion on $(\Omega, \mathcal{B}, P)$, then there is another Brownian motion $\{y_t\}$ on $(\Omega, \mathcal{B}, P)$ such that

(i) for each t, $x_t = y_t$ a. e.; and

(ii) for each ω, y_t is a continuous real-valued function of t.

Thus <u>there is no loss of generality in assuming that the Brownian paths $\{x_t(\omega) : t \geq 0\}$ are continuous.</u>

This observation also shows that Brownian motion, if it exists, is essentially unique. For if we identify the point $\omega \in \Omega$ with the Brownian history $y_t(\omega)$, so that $\omega(t) = y_t(\omega)$, then we have an inclusion $\Omega \subset \mathcal{C}[0, \infty)$, and the σ-algebra $\mathcal{B}$ and measure P are carried over in an obvious way. Because of the preceding remark, every Brownian motion is essentially equivalent to this one. Thus the problem of constructing the unique Brownian motion is that of finding a probability measure on $\mathcal{C}[0, \infty)$ which satisfies conditions (i), (ii), and (iii) of the definition of Brownian motion. The σ-algebra of measurable subsets of $\mathcal{C}[0, \infty)$ will be the one generated by all sets of the form $\{f : f(s + t) - f(t) \in A\}$, for $s, t \geq 0$ and A a Borel subset of $\mathbf{R}$. This σ-algebra is the same as the one generated by the point evaluations $\omega \to \omega(t) = x_t(\omega)$, and it coincides also with the Borel σ-algebra generated by the topology of uniform convergence on compacta [48, VI. 2].

3.2 Construction

We begin with a real-valued function $y = f(x)$ on $[-\frac{1}{2}, \frac{1}{2}]$ which has a Gaussian distribution, so that for each Borel subset E of $\mathbf{R}$,

$$m\{x : f(x) \in E\} = \frac{1}{\sqrt{(2\pi)}} \int_E e^{-u^2/2} du ,$$

where m denotes Lebesgue measure on $\mathbf{R}$. For example, the equation

$$x = \frac{1}{\sqrt{(2\pi)}} \int_0^y e^{-u^2/2} du$$

defines such a function $y = f(x)$.

Now let $\Omega = \prod_{n=1}^{\infty} [-\frac{1}{2}, \frac{1}{2}]$, and endow Ω with the product σ-algebra and the product measure P determined by the Borel field and Lebesgue measure on $[-\frac{1}{2}, \frac{1}{2}]$. For each $n = 1, 2, \ldots$ define a function $f_n : \Omega \to \mathbf{R}$ by $f_n(\omega) = f(\omega_n)$. It is clear that $\{f_n\}$ forms a family of independent random variables on Ω. Let $\mathcal{X}$ be the closed linear span in $L^2(\Omega, \mathcal{B}, P)$ of $\{f_n : n = 1, 2, \ldots\}$. (Note that a function with Gaussian distribution is in L^p for each p with $0 < p < \infty$.) It can be proved that every function in $\mathcal{X}$ has a Gaussian distribution, and the σ-algebra generated by $\mathcal{X}$ is the full product σ-algebra of Ω.

Because $\mathcal{X}$ is an infinite-dimensional separable Hilbert space, it is unitarily equivalent to $L^2[0, \infty)$. Let $U : L^2[0, \infty) \to \mathcal{X}$ be a unitary operator which realizes this equivalence. For each $t \geq 0$, let

$$x_t = U\chi_{[0, t]}$$

be the image under U of the characteristic function of the interval $[0, t]$.

(i) Clearly $x_0 = U(0) = 0$ a.e.

(ii) For $s, t \geq 0$, $x_{t+s} - x_t \in \mathcal{X}$ and hence $x_{t+s} - x_t$ has a Gaussian distribution with mean 0. Because U is unitary, we see that the variance of $x_{t+s} - x_t$ is

$$\|x_{t+s} - x_t\|_2^2 = \|\chi_{[0, t+s]} - \chi_{[0, t]}\|^2_{L^2(\mathbf{R})} = s.$$

(iii) If $0 \leq t_0 < t_1 < \ldots < t_n$, then the functions $\chi_{[0, t_{i+1}]} - \chi_{[0, t_i]}$ are pairwise orthogonal in $L^2(\mathbf{R})$, and hence their images $x_{t_{i+1}} - x_{t_i}$ are pairwise orthogonal in $L^2(\Omega)$. Now independent random variables with mean 0 are always orthogonal, and for Gaussian random variables with mean 0 the two concepts coincide. Therefore the increments $x_{t_{i+1}} - x_{t_i}$, $i = 0, \ldots, n - 1$, are independent.

This proves that $\{x_t : t \geq 0\}$ is Brownian motion.

3.3 Some properties of the Brownian paths

The ensemble of Brownian travelers moves according to subtle and remarkable principles. Some feeling for the nature of the process may be obtained from consideration of the following facts.

3.3.1 Unbounded variation. For almost all ω, $x_t(\omega)$ (as a function of t) does not have bounded variation on any interval.

3.3.2 Local law of the iterated logarithm. A more exact idea of the modulus of continuity of a typical Brownian path is given by this theorem of Khintchine [32] and Lévy [35, 36]:

For each $t \geq 0$, $P\{\omega : \limsup_{h \to 0^+} \frac{x_{t+h}(\omega) - x_t(\omega)}{\sqrt{(2h \log \log \frac{1}{h})}} = 1\} = 1.$

3.3.3 **Non-differentiability.** An immediate consequence of the local law of the iterated logarithm is that for a fixed t, and almost all ω, $x_t(\omega)$ is not differentiable at t. It can be proved that in fact $\chi_t(\omega)$ is nowhere differentiable for almost all ω. (And thus, in the sense of Wiener measure P on $\mathcal{C}[0, \infty)$, 'almost all' continuous functions are nowhere differentiable.)

3.3.4 **Law of the iterated logarithm.**

$$P\{\omega : \limsup_{t \to \infty} \frac{x_t(\omega)}{\sqrt{(2t \log \log t)}} = 1\} = 1$$

(Khintchine [31]).

This shows that the Brownian traveler wanders off to infinity, and gives an idea of the speed with which he does so. However, we will see momentarily that he still returns to any neighbourhood of the origin infinitely many times.

3.3.5 Brownian motion in R^n. On the n-fold product probability space $(\Omega \times \ldots \times \Omega, \mathcal{B} \times \ldots \times \mathcal{B}, P \times \ldots \times P)$ for $t \geq 0$ let $z_t(\omega) = (x_t(\omega_1), x_t(\omega_2), \ldots, x_t(\omega_n))$. Then $z_t(\omega)$, whose coordinates are independent one-dimensional Brownian motions, is n-dimensional Brownian motion.

The densities associated with these Brownian motions are easily computed. For example, if $n = 2$ and $U \subset R^2$ is measurable, then for $s, t \geq 0$

$$P\{\omega : z_{s+t}(\omega) - z_t(\omega) \in U\} = \iint_U \frac{1}{\sqrt{(2\pi s)}} e^{-x_1^2/2s} \frac{1}{\sqrt{(2\pi s)}} e^{-x_2^2/2s} dx_1 dx_2$$

$$= \frac{1}{2\pi s} \iint_U e^{-r^2/2s} r dr d\theta,$$

because of the independence of the two coordinate functions.

Some interesting properties of n-dimensional Brownian motion follow from the theorem of Kakutani mentioned in the Introduction (and proved below, in 3.6).

Recurrence: Brownian motion in dimensions 1 and 2 returns infinitely many times to each neighbourhood of the origin; but in dimensions $n \geq 3$, $|z_t(\omega)| \to \infty$ a.e. as $t \to \infty$.

Density: The paths of two-dimensional Brownian motion are everywhere dense in the plane, with probability 1; but for $n \geq 3$, the Brownian paths tend to infinity with increasing t.

Let us consider the proofs of these statements. Fix $n = 1, 2, \ldots$, $x_0 \in \mathbf{R}^n$, and r and R with $0 < r < R$, and let $E = \{x \in \mathbf{R}^n : r < |x - x_0| < R\}$. Let u be the solution in E of the Dirichlet problem with boundary values 1 on the inner sphere and 0 on the outer sphere. By Kakutani's theorem, for each $x \in E$ the number $u(x)$ is the probability that a Brownian traveler starting at x hits the inner sphere before the outer one. However, it is known that

$$u(x) = \frac{\log \frac{1}{|x-x_0|} - \log \frac{1}{R}}{\log \frac{1}{r} - \log \frac{1}{R}} \quad \text{if } n = 2, \text{ while}$$

$$u(x) = \frac{\frac{1}{|x-x_0|^{n-2}} - \frac{1}{R^{n-2}}}{\frac{1}{r^{n-2}} - \frac{1}{R^{n-2}}} \quad \text{if } n \geq 3.$$

Letting $R \to \infty$, we see that $u(x) \to 1$ if $n = 2$ while $u(x) \to (r/|x - x_0|)^{n-2} < 1$ if $n \geq 3$. Thus the probability that Brownian motion will enter a small sphere of radius r centered at x_0 is 1 if $n = 2$.

On the other hand, suppose that $n \geq 3$ and let A denote the set of those paths ω for which $\lim_{t\to\infty} |z_t(\omega)| \neq \infty$. Because of the law of the iterated logarithm, almost every path of A will contain points where $|z_t(\omega)|$ is arbitrarily large; and since $\{z_t\}$ forms a Markov process (see 3.4), the probability that a Brownian path whose modulus has exceeded a value K will decrease in modulus sometime again to a much smaller value k is no greater than $(k/K)^{n-2}$. By fixing $j \geq 1$ and letting $K = k^{j+1}$, we see that if A_k denotes the set of paths which return infinitely many times to a ball of radius k centered at the origin, then $P(A_k) \leq k^{-j(n-2)}$. Letting j tend to infinity shows that $P(A_k) = 0$, and hence $P(A) = P(\cup A_k) = 0$.

There are other interesting variations in the behavior of Brownian motion depending upon the dimension. For example, Dvoretzky, Erdős, and Kakutani [17] have proved that only when $n \leq 3$ is there a positive probability that the Brownian path will cross itself, and that then this probability equals 1.

3.3.6 **Conformal invariance.** Let $\gamma_{0,t}(\omega) = x_t(\omega) + iy_t(\omega)$ be Brownian motion in the complex plane starting at the origin. If F is a non-constant analytic function on D, then, up to the exit time from D, $\{F(\gamma_{0,t}) : t \geq 0\}$ is again Brownian motion, with respect to a new time parameter

$$T(t) = \int_0^t |F'(\gamma_{0,s})|^2 ds.$$

This theorem of Lévy [36] makes possible, for example, the use of martingale arguments about the transformed motion (see 5.1.2 and 5.2).

3.3.7 **The flow associated with Brownian motion.** Let $\{x_t : t \in \mathbf{R}\}$ be two-sided Brownian motion on $(\Omega, \mathcal{B}, P)$, and let $\mathcal{J}$ denote the family of all closed subintervals $[a, b]$ of $\mathbf{R}$. Each $I \in \mathcal{J}$ determines a Borel measure P_I on $\mathbf{R}$ according to the formula

$$P_I(E) = P\{\omega : x_I(\omega) \in E\},$$

where $x_I(\omega) = x_a(\omega) - x_b(\omega)$ if $I = [a, b]$. The family of translations

$$\phi_t : \mathbf{R}^{\mathcal{J}} \to \mathbf{R}^{\mathcal{J}} \qquad (t \geq 0)$$

defined by

$$(\phi_t f)(I) = f(I + t) \text{ for } I \in \mathcal{J} \text{ and } f \in \mathbf{R}^{\mathcal{J}}$$

is a <u>one-parameter flow</u> on $\mathbf{R}^{\mathcal{J}}$. Because of the stationarity of the Brownian increments, this flow preserves the product of the measures P_I on the product measure space $\mathbf{R}^{\mathcal{J}}$.

Many researchers have looked into the ergodic-theoretic properties of this flow. For example, Wiener [57] and Hopf [25] proved that it is strongly mixing; Anzai [1] uncovered several of its interesting aspects, including its behavior with respect to skew products; and Ornstein and

Shields [46] showed that if the Brownian motion is confined to a rectangular region by means of reflecting barriers, then the resulting flow is Bernoulli.

3.4 The martingale and Markov properties

The definitions of the terms 'martingale' and 'Markov process' depend on the notions of conditional expectation and conditional probability.

Let $(\Omega, \mathcal{B}, P)$ be a probability space, $\mathcal{A} \subset \mathcal{B}$ a sub-σ-algebra, and $f \in L^1(\Omega, \mathcal{B}, P)$. The measure

$$\nu(A) = \int_A f dP \qquad (A \in \mathcal{A})$$

on $\mathcal{A}$ is absolutely continuous with respect to $P|\mathcal{A}$, and so by the Radon-Nikodym Theorem there is an $\mathcal{A}$-measurable L^1 function g such that

$$\nu(A) = \int_A f dP = \int_A g dP \quad \text{for all } A \in \mathcal{A}.$$

Note that this equation determines g uniquely a.e. The function g is denoted by $E(f|\mathcal{A})$ and is called the <u>conditional expectation</u> of f with respect to $\mathcal{A}$.

$E(f|\mathcal{A})$ may be thought of as the expected value of the random variable f given the information contained in the σ-algebra $\mathcal{A}$; that is, if we know for each set of $\mathcal{A}$ whether or not it contains ω, then $E(f|\mathcal{A})(\omega)$ is the expected value of f on the basis of this knowledge.

The conditional expectation operator $E(\cdot|\cdot)$ has the following properties:

(i) If $f \geq 0$ a.e., then $E(f|\mathcal{A}) \geq 0$ a.e.

(ii) For each $p \geq 1$, $E(\cdot|\mathcal{A})$ is a linear operator of norm 1 on $L^p(\Omega, \mathcal{B}, P)$.

(iii) If f is $\mathcal{A}$-measurable, then $E(f|\mathcal{A}) = f$ a.e.

(iv) If f is $\mathcal{A}$-measurable, then $E(fg|\mathcal{A}) = fE(g|\mathcal{A})$ a.e.

(v) $E(f) = E(E(f|\mathcal{A}))$ $(E(f) = \int_\Omega f dP)$.

(vi) If $\mathcal{A}_0$ is a sub-σ-algebra of $\mathcal{A}$, then $E(E(f|\mathcal{A})|\mathcal{A}_0) = E(f|\mathcal{A}_0)$ a.e.

(vii) If f is independent of $\mathcal{A}$, then $E(f|\mathcal{A}) = E(f)$ a.e.

(viii) If $\phi : \mathbf{R} \to \mathbf{R}$ is convex (so that whenever $\alpha_1, \ldots, \alpha_n$ are nonnegative with sum 1, and $x_1, \ldots, x_n \in \mathbf{R}$, $\phi\sum\alpha_i x_i \leq \sum\alpha_i\phi(x_i)$), then $\phi(E(f|\mathcal{A})) \leq E(\phi \circ f|\mathcal{A})$ a.e.

If $f_1, f_2, \ldots$ are measurable functions on Ω, then we denote by $\mathcal{B}(f_1, f_2, \ldots)$ the smallest sub-σ-algebra of $\mathcal{B}$ with respect to which $f_1, f_2, \ldots$ are all measurable, and we abbreviate $E(f|\mathcal{B}(f_1, f_2, \ldots))$ by $E(f|f_1, f_2, \ldots)$.

If $E \in \mathcal{B}$, the conditional probability of E given $\mathcal{A}$ is defined to be

$$P(E|\mathcal{A}) = E(\chi_E|\mathcal{A}).$$

Note that if $E, F \in \mathcal{B}$, neither $P(F)$ nor $P(\Omega\backslash F)$ is 0, and $\mathcal{F} = \{\emptyset, \Omega, F, \Omega\backslash F\}$, then

$$P(E|\mathcal{F}) = \begin{cases} P(E|F) & \text{on } F \\ P(E|\Omega\backslash F) & \text{on } \Omega\backslash F, \end{cases}$$

so that this definition is related to the elementary concept of conditional probability.

Let $(\Omega, \mathcal{B}, P)$ be a probability space, T an ordered set, and $\{\mathcal{A}_t\}$ a family of sub-σ-algebras of $\mathcal{B}$ indexed by the elements of T. We suppose that $\{\mathcal{A}_t\}$ is increasing, so that $s \le t$ implies $\mathcal{A}_s \subset \mathcal{A}_t$. A family of (real or complex-valued) measurable functions $\{x_t : t \in T\}$ is called a stochastic process. A stochastic process $\{x_t : t \in T\}$ is called a martingale with respect to $\{\mathcal{A}_t\}$ in case

(i) $x_t \in L^1(\Omega, \mathcal{B}, P)$ for all t,

(ii) x_t is $\mathcal{A}_t$-measurable for all t, and

(iii) if $s < t$, then $E(x_t|\mathcal{A}_s) = x_s$ a. e.

A stochastic process $\{x_t : t \in T\}$ is called a martingale in case it is a martingale with respect to the σ-algebras $\mathcal{A}_t = \mathcal{B}\{x_s : s \le t\}$. In this case condition (iii) may be replaced by

(iii') if $t_1 < \ldots < t_{n+1}$, then $E(x_{t_{n+1}}|x_{t_1}, \ldots, x_{t_n}) = x_{t_n}$ a. e.

It may be helpful to think of the martingale condition (iii) in the following terms. If x_t is the fortune at time t of a gambler, this condition says that the game he is playing is a fair one: his expected fortune at any future time, given some of the past history of his play, is equal to his current holdings. If condition (iii) is replaced by

(iii") if $s < t$, then $E(x_t|\mathcal{A}_s) \ge x_s$ a. e., so that the game is advantageous to the gambler, then we say that $\{x_t\}$ is a submartingale with respect to $\{\mathcal{A}_t\}$. If $p \geq 1$ and $\{x_t\}$ is a martingale with respect to $\{\mathcal{A}_t\}$, then $\{|x_t|^p\}$ is a submartingale with respect to $\{\mathcal{A}_t\}$ [14, p. 296].

Most of the basic theory of martingales and submartingales is due to Doob. The following theorem is an example of the important properties that are shared by these kinds of stochastic processes.

3.4.1 **Submartingale convergence theorem.** Let $\{x_1, x_2, \ldots\}$ be a submartingale, and suppose that $\sup_n E|x_n| < \infty$. Then $\{x_n\}$ converges a.e. to a function x_∞, and $E|x_\infty| < \infty$. If $x_1, x_2, \ldots$ are uniformly integrable, in the sense that

$$\lim_{n\to\infty} \int_{\{\omega : |x_t(\omega)| > n\}} |x_t| dP = 0 \text{ uniformly in } t,$$

then $\{x_n\}$ converges to x_∞ in L^1.

3.4.2 **The martingale property.** The Brownian motion process $\{x_t : t \geq 0\}$ forms a martingale. In fact, for each fixed $t_0 \geq 0$, $\{x_t - x_{t_0} : t \geq t_0\}$ is a martingale. (However, $E|x_t|$ is unbounded, so there is no limit random variable.)

A stochastic process $\{x_t : t \in T\}$ is called a Markov process in case for each measurable set U and $t_1 < \ldots < t_{n+1}$,

$$P(x_{t_{n+1}} \in U | x_{t_1}, \ldots, x_{t_n}) = P(x_{t_{n+1}} \in U | x_{t_n}) \text{ a.e.}$$

Equivalent conditions are that for each t and each $f \in L^1(\Omega)$ measurable with respect to $\mathcal{B}\{x_s : s \geq t\}$,

$$E(f | x_s \text{ for } s \leq t) = E(f | x_t) \text{ a.e.};$$

or that for each measurable function f on $\mathbf{R}$ and $t_1 < \ldots < t_{n+1}$,

$$E(f(x_{t_{n+1}}) | x_{t_1}, \ldots, x_{t_n}) = E(f(x_{t_{n+1}}) | x_{t_n}) \text{ a.e.}$$

In any case, the idea behind the definition is that a Markov process has no memory: at each instant, the probability that the process will move on to a certain locale depends only on the present location of the process and not in any way on its previous history.

3.4.3 **The Markov property.** For each fixed $t_0 \geq 0$, the Brownian motion increments $\{x_t - x_{t_0} : t \geq t_0\}$ form a Markov process.

Even more, the Brownian motion process begins again at every instant: for $t \geq t_0$, $x_t - x_{t_0}$ is independent of $\mathcal{B}\{x_s : s \leq t_0\}$ and has the same distribution as x_{t-t_0}.

3.5 The strong Markov property

Not only does the Brownian motion start anew at every instant, but the instant chosen may be varied with the particular path, so long as this is done in a sufficiently regular way. The precise delineation of this 'strong Markov property' was accomplished by Hunt [26] and Blumenthal [3].

Let $\{x_t : t \geq 0\}$ be one-dimensional Brownian motion, and denote by $\mathcal{B}_t$ the σ-algebra generated by $\{x_s : 0 \leq s \leq t\}$. A measurable function $\tau : \Omega \to [0, \infty]$ is called a stopping time for $\{x_t\}$ if for each $s > 0$

$$\{\omega : \tau(\omega) < s\} \in \mathcal{B}_s.$$

A typical example of a stopping time is the first hitting time $\inf\{t : x_t \in A\}$ of a Borel set $A \subset \mathbf{R}$.

More generally, let $\{\mathcal{A}_t : t \geq 0\}$ be a family of sub-σ-algebras of $\mathcal{B}$ such that

(i) if $s < t$ then $\mathcal{A}_s \subset \mathcal{A}_t$,

(ii) $\mathcal{B}_t \subset \mathcal{A}_t$ for all t, and

(iii) for $t_0 > 0$, $\mathcal{A}_{t_0}$ is independent of $\{x_t - x_{t_0} : t \geq t_0\}$.

(One way to manufacture such a family is to start with a σ-algebra $\mathcal{A}$ which is independent of $\{x_t : t \geq 0\}$ and let $\mathcal{A}_t = \mathcal{A} \vee \mathcal{B}_t$.) A measurable function $\xi : \Omega \to [0, \infty]$ is called an $\mathcal{A}_t$-stopping time if for each $s > 0$

$$\{\omega : \xi(\omega) < s\} \in \mathcal{A}_s.$$

Let ξ be an $\mathcal{A}_t$-stopping time. The pre-ξ σ-algebra $\mathcal{A}_{\xi^+}$ is defined to be

$$\mathcal{A}_{\xi^+} = \{E \in \mathcal{B} : E \cap \{\omega : \xi(\omega) < s\} \in \mathcal{A}_s \text{ for each } s \geq 0\}.$$

The σ-algebra $\mathcal{A}_{\xi^+}$, which is defined as a sort of limit of the $\mathcal{A}_s$ as s decreases to ξ, is intended to accumulate all the information which is in the σ-algebras $\mathcal{A}_s$ so long as $s \leq \xi$. (Note that if ξ is constant then $\mathcal{A}_{\xi^+} = \bigcap_{s>\xi} \mathcal{A}_s$.) We may think of $\mathcal{A}_{\xi^+}$, then, as consisting of all those events which depend only on the happenings in the $\mathcal{A}_s$ up to time ξ. The observation that $x_{\xi(\omega)}(\omega)$ is measurable with respect to $\mathcal{A}_{\xi^+}$ may convey some idea of the size of this σ-algebra.

Theorem 3.5.1. _Let $\{\mathcal{A}_t : t \geq 0\}$ be as above and suppose that ξ is an $\mathcal{A}_t$-stopping time. Let $(\Omega_\xi, \mathcal{B}_\xi, P_\xi)$ be the probability subspace of $(\Omega, \mathcal{B}, P)$ determined by $\Omega_\xi = \{\omega : \xi(\omega) < \infty\}$, $\mathcal{B}_\xi = \{A \cap \Omega_\xi : A \in \mathcal{B}\}$, and $P(A) = P(A | \Omega_\xi)$. Then_

(i) $\{x_{t+\xi(\omega)}(\omega) - x_{\xi(\omega)}(\omega) : t \geq 0\}$ _is Brownian motion on $(\Omega_\xi, \mathcal{B}_\xi, P_\xi)$, and_

(ii) _this process is independent of $\Omega_\xi \cap \mathcal{A}_{\xi^+}$ (in $(\Omega_\xi, \mathcal{B}_\xi, P_\xi)$)._

Of course a similar statement holds for n-dimensional Brownian motion.

An easily applicable consequence of this strong Markov property can be stated in terms of conditional probabilities, and in the following section we will use this formulation to sketch a proof of the theorem of Kakutani on the Dirichlet problem. The argument is typical of several which will appear in the following pages. First we need a lemma on independent random variables.

Lemma 3.5.2. _If X_1 and X_2 are independent random variables on a probability space $(\Omega, \mathcal{B}, P)$, $A \subset \Omega$ is measurable, X_2 is independent of the σ-algebra generated by A and X_1, and $f : \mathbf{R}^2 \to \mathbf{R}$ is measurable, then_

$$\int_{\omega_1 \in A} \int_{\omega_2 \in \Omega} f(X_1(\omega_1), X_2(\omega_2)) dP(\omega_1) dP(\omega_2) = \int_A f(X_1(\omega), X_2(\omega)) dP(\omega).$$

Proof. For $i = 1, 2$ and measurable $E \subset \mathbf{R}$ let $P_{X_i}(E) = P\{\omega : X_i(\omega) \in E\}$ and $P_{X_i|_A}(E) = P\{\omega \in A : X_i(\omega) \in E\}$. For measurable subsets E and F of $\mathbf{R}$, define $P_{X_1|_A, X_2|_A}(E \times F) = P\{\omega : (X_1|_A(\omega), X_2|_A(\omega)) \in E \times F\}$; then

$$P_{X_1|_A, X_2|_A}(E \times F) = P\{\omega \in A : X_1(\omega) \in E, X_2(\omega) \in F\}.$$

$$= P\{\omega : \omega \in A, X_1(\omega) \in E, X_2(\omega) \in F\} = P_{X_1|_A}(E)P_{X_2}(F).$$

Therefore, by two changes of variables,

$$\iint_{A\times\Omega} f(X_1(\omega_1), X_2(\omega_2))dP(\omega_1)dP(\omega_2) = \iint_{R^2} f(x, y)dP_{X_1|_A}(x)dP_{X_2}(y)$$

$$= \int_{R^2} f(x, y)dP_{X_1|_A, X_2|_A}(x, y) = \int_A f(X_1(\omega), X_2(\omega))dP(\omega).$$

Let $\{x_t : t \geq 0\}$ be a Brownian motion process. For each fixed x, $\{x + x_t : t \geq 0\}$ is Brownian motion starting at x, and we denote by P_x the probability associated with this process: for example, for measurable $E \subset \mathbf{R}$, $P_x\{x_t \in E\} = P\{\omega : x + x_t(\omega) \in E\}$.

Theorem 3.5.3. Let $\{\mathcal{A}_t : t \geq 0\}$ be a family of sub-σ-algebras of $\mathcal{B}$ satisfying (i), (ii), and (iii) above (p. 26), and let ξ be an $\mathcal{A}_t$-stopping time. Then for each measurable $E \subset \mathbf{R}$ and each $t \geq 0$,

$$P(x_{t+\xi} \in E | \mathcal{A}_{\xi^+}) = P_{x_\xi}\{x_t \in E\} \text{ a.e. on } \{\omega : \xi(\omega) < \infty\}.$$

Proof. Each side of the above equation is a function of ω, the right side becoming

$$P_{x_{\xi(\omega)}(\omega)}\{x_t \in E\}$$

when ω is supplied. Of course the left side is measurable with respect to $\mathcal{A}_{\xi^+}$; and, since x_ξ is measurable with respect to $\mathcal{A}_{\xi^+}$, so is $P_{x_\xi}\{x_t \in E\}$. Therefore, because of the essential uniqueness of conditional probability, it is enough to show that

$$\int_A P_{x_{\xi(\omega)}(\omega)}\{x_t \in E\}dP(\omega) = \int_A \chi_E(x_{t+\xi(\omega)}(\omega))dP(\omega)$$

for all $A \in \mathcal{A}_{\xi^+}$.

By the strong Markov property, $x_{t+\xi(\omega')}(\omega') - x_{\xi(\omega')}(\omega')$ is independent of $A \in \mathcal{A}_{\xi^+}$ and $x_{\xi(\omega)}(\omega)$ (which is measurable with respect to $\mathcal{A}_{\xi^+}$) and has the same distribution as $x_t(\omega')$. Therefore the Lemma applies with $f(x, y) = \chi_E(x + y)$, and

$$\int_A P_{x_{\xi(\omega)}(\omega)}\{x_t \in E\}dP(\omega) = \int_A P\{\omega' : x_{\xi(\omega)}(\omega)+x_t(\omega')\in E\}dP(\omega)$$

$$= \int_{A\times\Omega} \chi_E(x_{\xi(\omega)}(\omega) + x_t(\omega'))dP(\omega')dP(\omega)$$

$$= \int_{A\times\Omega}\chi_E(x_{\xi(\omega)}(\omega) + x_{t+\xi(\omega')}(\omega') - x_{\xi(\omega)}(\omega'))dP(\omega')dP(\omega)$$

$$= \int_A \chi_E(x_{\xi(\omega)}(\omega) + x_{t+\xi(\omega)}(\omega) - x_{\xi(\omega)}(\omega))dP(\omega)$$

$$= \int_A \chi_E(x_{t+\xi(\omega)}(\omega))dP(\omega) \quad \text{for all } A \in \mathcal{A}_{\xi^+} .$$

Corollary 3.5.4. <u>Let</u> $\{\gamma_{z,t}(\omega) : t \geq 0\}$ <u>be n-dimensional Brownian motion starting at</u> $z \in \mathbf{R}^n$, $\mathcal{B}_t = \mathcal{B}\{\gamma_{0,s} : 0 \leq s \leq t\}$, ξ <u>a</u> $\mathcal{B}_t$<u>-stopping time, and</u> $Y(\omega) = \gamma_{z,\,\xi(\omega)}(\omega)$. <u>If</u> g <u>is a measurable function on</u> $\mathbf{R}^n$ <u>such that</u> $g(\gamma_{z,t}) \in L^1(\Omega)$ <u>for all</u> z <u>and</u> t, <u>and if</u> $t \geq 0$ <u>and</u> $z \in \mathbf{R}^n$, <u>then</u>

$$E(g(\gamma_{z,\,t+\xi})|\mathcal{B}_{\xi^+})(\omega) = E(g(\gamma_{Y(\omega),\,t})) \quad \text{a. e.}$$

Proof. According to the Theorem,

$$E(\chi_E(\gamma_{z,\,t+\xi})|\mathcal{B}_{\xi^+})(\omega) = E(\chi_E(\gamma_{Y(\omega),\,t})) \quad \text{a. e. } dP(\omega)$$

for each measurable $E \subset \mathbf{R}^n$. Because of linearity, a similar formula holds when χ_E is replaced by a linear combination of characteristic functions, and the result follows from the Dominated Convergence Theorem.

3.6 Kakutani's theorem

The following theorem is essential for the arguments to be encountered in what follows, so we will examine its proof in some detail.

Theorem 3.6.1 (Kakutani [29]). Let S be a bounded, connected, open subset of R^n with smooth boundary ∂S, f a continuous function on ∂S, $\gamma_{z,t}(\omega)$ an n-dimensional Brownian motion which starts at $z \in S$, and $T_z(\omega) = \inf\{t \geq 0 : \gamma_{z,t}(\omega) \notin S\}$. Then

$$u(z) = \int_\Omega f(\gamma_{z, T_z(\omega)}(\omega))dP(\omega)$$

is harmonic on S and $\lim_{z \to x} u(z) = f(x)$ for each $x \in \partial S$.

Proof. We will show that the bounded function u is harmonic by proving that for every $z \in S$, $u(z)$ equals the average value of u over each small sphere centered at z. Fix $z \in S$, let Σ be a sphere centered at z which, together with its interior, is entirely contained in S, let $\xi(\omega) = \inf\{t \geq 0 : \gamma_{z,t}(\omega) \in \Sigma\}$ be the first hitting time of $\gamma_{z,t}$ on Σ, and let $Y(\omega) = \gamma_{z, \xi(\omega)}(\omega)$ be the first point of impact of $\gamma_{z,t}$ on Σ. The strong Markov property allows us to break up each Brownian path from z to ∂S into two independent pieces, one to the first impact on Σ and a subsequent one to ∂S. Moreover, because any rotation of Brownian motion is again Brownian motion, Y is uniformly distributed over Σ. Therefore we may try to write

$$u(z) = \int_\Omega f(\gamma_{z,T_z})dP = \int_\Omega \int_\Sigma f(\gamma_{\zeta, T_\zeta})P\{\gamma_{z,\xi} \in d\zeta\}dP$$

$$= \int_\Sigma u(\zeta)P\{Y \in d\zeta\} = \int_\Sigma u(\zeta)dm(\zeta),$$

where m denotes normalized Lebesgue measure on Σ. Of course such a calculation needs a bit more explanation and support before it can be accepted.

First, the rotational invariance of the densities associated with n-dimensional Brownian motion (see 3.3.5) shows that the Borel measure μ_Y on Σ defined by $\mu_Y(E) = P\{\omega : Y(\omega) \in E\}$ is invariant under rotations and therefore coincides with m. Since two of the integrals above are to be understood as taken with respect to μ_Y -

$$\int_\Sigma f(\gamma_{\zeta, T_\zeta}) P\{\gamma_{z,\xi} \in d\zeta\} = \int_\Sigma f(\gamma_{\zeta, T_\zeta}) d\mu_Y(\zeta) \text{ and}$$

$$\int_\Sigma u(\zeta) P\{Y \in d\zeta\} = \int_\Sigma u(\zeta) d\mu_Y(\zeta) \text{ -}$$

only the first integral equality remains to be verified.

At this point it is convenient to replace our Brownian motion $\gamma_{z,t}(\omega)$ by a new stochastic process

$$\tilde{\gamma}_{z,t}(\omega) = \gamma_{z, t\wedge T_z(\omega)}(\omega)$$

which has ∂S as an <u>absorbing barrier</u> ($\wedge$ denotes infimum). This process inherits from $\gamma_{z,t}$ the conditional-probability version of the strong Markov property [18, I, Ch. 10], and the preceding Corollary applies also to $\{\tilde{\gamma}_{z,t} : t \geq 0\}$.

Extend the domain of f by letting $f(s) = 0$ for all $s \in S$. Then for $t \geq 0$,

$$\int_\Omega f(\tilde{\gamma}_{z,t+\xi}) dP = \int_\Omega E(f(\tilde{\gamma}_{z,t+\xi}) | \mathcal{B}_{\xi+}) dP$$

$$= \int_\Omega E(f(\tilde{\gamma}_{Y(\omega),t})) dP = \int_\Omega \int_\Omega f(\tilde{\gamma}_{Y(\omega),t}(\omega')) dP(\omega') dP(\omega)$$

$$= \int_\Omega \int_\Sigma f(\tilde{\gamma}_{\zeta,t}(\omega')) d\mu_Y(\zeta) dP(\omega'),$$

and taking limits as $t \to \infty$ gives

$$\int_\Omega f(\gamma_{z,T_z}) dP = \int_\Omega \int_\Sigma f(\gamma_{\zeta, T_\zeta}) d\mu_Y(\zeta) dP,$$

as required.

The first step in proving that $\lim_{z\to x} u(z) = f(x)$ for each $x \in \partial S$ is to show that for every small neighbourhood N of $x \in \partial S$, $P\{\gamma_{z,T_z} \in N\}$ tends to 1 as z approaches x. Suppose then that $x \in \partial S$ is given and that N is the intersection of a small ball B centered at x_0 with ∂S. Choose an orthogonal coordinate system $(x^1, \ldots, x^n)$ for $\mathbf{R}^n$ in such a way that the x^1 axis is perpendicular to the tangent plane to ∂S at x_0.

Because of the rotational invariance of Brownian motion, we may assume that

$$\gamma_{z,t} = z + (x_t^1, \ldots, x_t^n),$$

where the x_t^i are independent one-dimensional Brownian motions for $i = 1, \ldots, n$.

For each $t \geq 0$ and $i = 2, \ldots, n$ let

$$F_t^i = \left\{ \omega : \sup_{0<h<t} \frac{x_h^i(\omega)}{\sqrt{(2h \log\log \frac{1}{h})}} < 2 \right\}.$$

Then the sets F_t^i increase as t decreases toward 0, and

$$\bigcup_{t>0} F_t^i = \left\{ \omega : \sup_{0<h<t} \frac{x_h^i(\omega)}{\sqrt{(2h \log\log \frac{1}{h})}} < 2 \text{ for some } t \geq 0 \right\}$$

$$= \left\{ \omega : \inf_{t>0} \sup_{0<h<t} \frac{x_h^i(\omega)}{\sqrt{(2h \log\log \frac{1}{h})}} < 2 \right\},$$

which, because of the local law of the iterated logarithm (3. 3. 2) has measure 1. It is possible, therefore, to choose t so close to 0 that $2\sqrt{(n-1)}\sqrt{(2t \log\log \frac{1}{t})}$ is less than the radius of B and, simultaneously, the intersection of the F_t^i, $i = 2, \ldots, n$, has probability very near 1.

If for this fixed value of t and $k = 1, 2, \ldots$ we let

$$E_{k,t} = \{\omega : \sup_{0<h<t} x_h^1(\omega) \geq \frac{1}{k}\},$$

then the $E_{k,t}$ increase with k and their union has measure 1. Therefore we may choose k so large that $E_{k,t}$ has probability very near 1.

Suppose now that to begin with N were chosen small enough that the accompanying figure (in which the cylinder C is swept out by translating the intersection of B with the tangent plane in the direction of the x^1 axis a distance $\frac{1}{2k}$ to each side of the tangent plane) represents the situation accurately. Our preceding discussion shows that if $z \in C \cap S$, then $\gamma_{z,T_z} \in N$ with a probability very near 1 (for each $\omega \in E_{k,t} \cap \bigcap_{i=2}^{n} F_t^i$, $\gamma_{z,T_z}(\omega) \in N$). Therefore $\lim_{z \to x} P\{\gamma_{z,T_z} \in N\} = 1$.

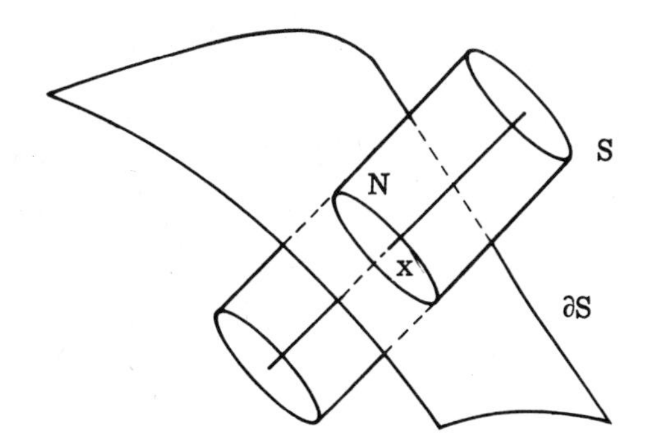

Because f is continuous on ∂S, this result implies immediately that $f(\gamma_{z, T_z})$ converges to the constant function f(x) in measure as $z \to x$, and hence $u(z) \to f(x)$.

Corollary 3.6.2. <u>Let</u> $\{\gamma_{z,t} : t \geq 0\}$ <u>denote Brownian motion in the complex plane starting at the point</u> z <u>at time</u> $t = 0$. <u>For each</u> $z \in D = \{\zeta : |\zeta| < 1\}$, <u>let</u> $\tau_z(\omega) = \inf\{t \geq 0 : |\gamma_{z,t}(\omega)| = 1\}$. <u>If</u> $E \subset \Gamma = \{\zeta : |\zeta| = 1\}$ <u>is measurable, then the hitting probability of</u> $\gamma_{z,t}$ <u>in</u> E <u>is the harmonic measure of</u> E <u>at</u> z:

$$P\{\omega : \gamma_{z, \tau_z(\omega)}(\omega) \in E\} = \int_E \mathcal{P}(r, \theta - t)dm(t)$$

(<u>where</u> $\mathcal{P}(r, \phi) = \dfrac{1 - r^2}{1 - 2r \cos \phi + r^2}$ <u>is the Poisson kernel and</u> $z = re^{i\theta}$).

Proof. Let f be continuous on Γ. A change of variables with $Y_z(\omega) = \gamma_{z, \tau_z(\omega)}$ and $\mu_z(A) = P\{\omega : Y_z(\omega) \in A\}$ (for measurable $A \subset \Gamma$) shows that

$$\int_\Omega f(\gamma_{z, \tau_z(\omega)}(\omega))dP(\omega) = \int_\Gamma f(\zeta)d\mu_z(\zeta).$$

By the Theorem, this is a harmonic function in D with boundary values f;

therefore, by classical results, it equals

$$\int_{\Gamma} f(e^{it}) \mathcal{P}(r, \theta - t) dm(t) .$$

Since this is true for all continuous functions f, $d\mu_z(e^{it}) = \mathcal{P}(r, \theta - t)dm(t)$, and, in particular,

$$P\{\omega : \gamma_{z, \tau_z(\omega)}(\omega) \in E\} = \mu_z(E) = \int_E \mathcal{P}(r, \theta - t)dm(t).$$

4 · Distribution equivalence of the two maximal functions

From now on we deal with two-dimensional Brownian motion in the unit disk $D = \{z : |z| < 1\}$ of the complex plane C : $\gamma_{z,t}(\omega)$ denotes the position at time $t \geq 0$ of the Brownian motion which at time 0 is at $z \in D$. Let

$$\tau_z(\omega) = \inf\{t \geq 0 : |\gamma_{z,t}(\omega)| = 1\}$$

be the first hitting time of $\gamma_{z,t}$ on $\Gamma = \partial D$. Sometimes we will denote $\gamma_{0,t}$ and τ_0 simply by γ_t and τ, respectively. For any function f on D and $\sigma \in (0, 1)$, recall that the Brownian and classical maximal functions are defined by

$$f^*(\omega) = \sup\{|f(\gamma_t(\omega))| : 0 \leq t < \tau(\omega)\} \quad \text{and}$$
$$N_\sigma f(e^{i\theta}) = \sup\{|f(z)| : z \in \Omega_\sigma(\theta)\},$$

where $\Omega_\sigma(\theta)$ is the 'Stolz domain' shown in the figure:

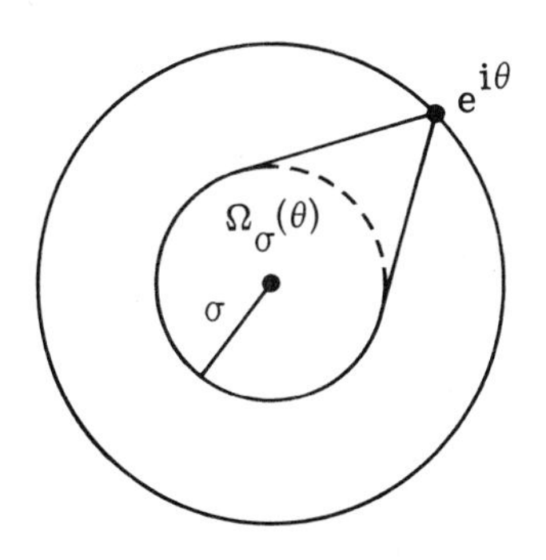

The goal of this and the following two chapters is the proof of the following theorem of Burkholder, Gundy, and Silverstein [7]:

Theorem 6.1. Let u be harmonic on D and $F = u + i\tilde{u}$ its analytic completion. Then for $0 < p < \infty$

$$c_{\sigma,p} \|N_\sigma u\|_{L^p(\Gamma)} \leq \|F\|_{H^p} \leq C_{\sigma,p} \|N_\sigma u\|_{L^p(\Gamma)} .$$

The probabilistic proof of this result is accomplished by carrying over the analysis to Ω, where martingale inequalities, stopping time arguments, and special properties of the Brownian paths can be used to advantage. The following theorem, which links the distributions of the Brownian and classical maximal functions, provides the mechanism for this transferral: the fact that $u^* \in L^p(\Omega)$ if and only if $N_\sigma u \in L^p(\Gamma)$.

Theorem 4.1. For each σ with $0 < \sigma < 1$ there are constants c_σ and C_σ such that whenever u is harmonic on D and $\lambda > 0$,

$$c_\sigma m\{e^{i\theta} : N_\sigma u(e^{i\theta}) > \lambda\} \leq P\{\omega : u^*(\omega) \geq \lambda\} \leq C_\sigma m\{e^{i\theta} : N_\sigma u(e^{i\theta}) > \lambda\}.$$

The right-hand inequality is proved fairly easily in the following section, with the help of the strong Markov property. The remaining three parts of the chapter are taken up by the proof of the left-hand inequality, which requires a fairly careful study of some conditional Brownian motions along with moderately accurate estimates of harmonic measures. The idea is to note first that if $\tilde{\gamma}_t = \gamma_{t\wedge\tau}$ never hits $S_r = \Omega_\sigma(\theta) \cap \{z : |z| = r\}$, and if u achieves a value larger than λ on S_r, then $\{\tilde{\gamma}_t(\omega) : t \geq 0\}$ together with its reflection $\{\hat{\gamma}_t(\omega) : t \geq 0\}$ about the segment from 0 to $e^{i\theta}$ contains a closed curve around S_r on which, because of the Maximum Principle, u must also achieve values larger than λ. In 4.2 we discuss the probabilities P^θ associated with the Brownian motion 'conditioned to hit Γ at $e^{i\theta}$', and in 4.4 we show that these functions are the same for the $\hat{\gamma}_t$ process as for $\tilde{\gamma}_t$. Therefore, if

$$E = \{\omega : \sup_{t \geq 0} |u(\tilde{\gamma}_t(\omega))| > \lambda\} \text{ and}$$

$$\hat{E} = \{\omega : \sup_{t \geq 0} |u(\hat{\gamma}_t(\omega))| > \lambda\},$$

it will follow that

$$P^\theta(E) = \tfrac{1}{2}(P^\theta(E) + P^\theta(\hat{E}))$$

$$\geq \tfrac{1}{2}P^\theta\{\omega : \sup_{z \in \tilde{\gamma}_t(\omega) \cup \hat{\gamma}_t(\omega)} |u(z)| > \lambda\}$$

$$\geq \tfrac{1}{2}P^\theta\{\omega : \tilde{\gamma}_t(\omega) \notin S_r \text{ for } t \geq 0\} \text{ for each } r,$$

and in 4.2 we show that for small σ and $r \geq \max\{\sigma, \frac{1}{2}\}$ the latter quantity (which is related to a certain harmonic measure) is no less than $c_\sigma > 0$. Denote by K the collection of all those $e^{i\theta} \in \Gamma$ for which $\sup\{|u(z)| : z \in \Omega_\sigma(\theta), |z| \geq \max\{\sigma, \frac{1}{2}\}\} > \lambda$. Then by a change of variables,

$$P\{\omega : u^*(\omega) > \lambda\} = \int_\Omega P(u^* > \lambda | \gamma_T) dP$$

$$= \int_\Gamma P^\theta\{\omega : u^*(\omega) > \lambda\} dm(\theta) \geq \int_K P^\theta(E) dm(\theta) \geq c_\sigma m(K).$$

Section 4.3 shows that $m(K) \geq c'_\sigma m\{e^{i\theta} : N_\sigma u(e^{i\theta}) > \lambda\}$ and removes the restriction on σ, so that this calculation will indeed complete the proof.

4.1 The right-hand inequality

Fix σ with $0 < \sigma < 1$ and $\lambda > 0$, and let

$$A = \{e^{i\theta} : N_\sigma u(e^{i\theta}) > \lambda\},$$

$$A' = \Gamma \backslash A,$$

$$B = \cup\{\Omega_\sigma(\theta) : e^{i\theta} \in A'\}, \text{ and}$$

$$E = \{z : |u(z)| > \lambda\}.$$

Assume that $A, A' \neq \emptyset$, the proof being trivial if this is not so. The boundary of B consists of parts of Γ and the circle $\{z : |z| = \sigma\}$ together with a saw-toothed curve in D:

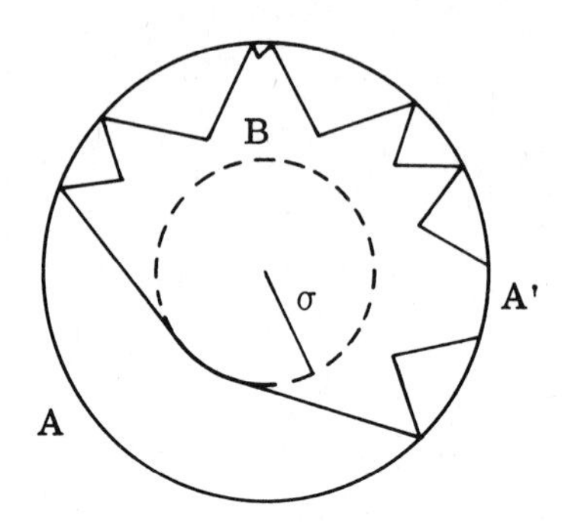

Notice that if $\gamma_t(\omega)$ never hits $\partial B \setminus \Gamma$ for $0 \leq t \leq \tau(\omega)$, then its first point of impact $\gamma_{\tau(\omega)}(\omega)$ with Γ is in A', and $\gamma_t(\omega)$ does not enter E before hitting Γ. Therefore

$$(1) \qquad P\{\omega : u^*(\omega) > \lambda\} = P\{\omega : \gamma_t(\omega) \in E \text{ for some } t \geq 0\}$$
$$\leq P\{\omega : \gamma_t(\omega) \in \partial B \setminus \Gamma \text{ for some } t \geq 0\}.$$

Assume that $\partial B \setminus \Gamma \neq \emptyset$ and fix $z = re^{i\theta} \in \partial B \setminus \Gamma$. By Corollary 3.11.5,

$$P\{\omega : \gamma_{z,\tau_z(\omega)}(\omega) \in A\} = \int_A \mathcal{P}(r, \theta - t)dm(t).$$

The first objective is to prove

$$(2) \qquad P\{\omega : \gamma_{z,\tau_z(\omega)}(\omega) \in A\} \geq \frac{\sigma}{\pi}.$$

Since $z \in \partial B \setminus \Gamma$, there is an entire subarc S of Γ, say from $e^{i\alpha}$ to $e^{i\beta}$, which is contained in A. Let S' be the projection of S onto Γ through z, as shown in the next figure. Clearly $e^{i\alpha}$ and $e^{i\beta}$ can be chosen so that $2\pi m(S') \geq 2\sigma$, even in case $|z| = \sigma$. By elementary geometry, the angle ϕ between the segments from $e^{i\alpha}$ and $e^{i\beta}$ to z is the average of the lengths of S and S'; then, because of the standard calculation [59, I, p. 99]

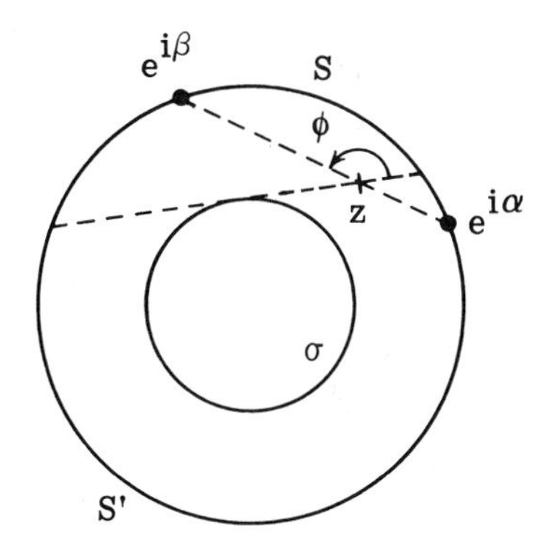

$$\pi[m(S) + m(S')] = \phi = \operatorname{Im} \int_{e^{i\alpha}}^{e^{i\beta}} \frac{d\zeta}{\zeta - z}$$

$$= \operatorname{Im} \int_{\alpha}^{\beta} \frac{ie^{it}dt}{e^{it} - re^{i\theta}} = \operatorname{Re} \int_{\alpha}^{\beta} \frac{dt}{1 - re^{i(\theta-t)}}$$

$$= \operatorname{Re} \int_{\alpha}^{\beta} \frac{1 - re^{-i(\theta-t)}}{1 - 2r\cos(\theta-t) + r^2} dt = \int_{\alpha}^{\beta} \frac{1 - r\cos(\theta - t)}{1 - 2r\cos(\theta-t) + r^2} dt$$

$$= \int_{\alpha}^{\beta} \frac{[\frac{1}{2} - r\cos(\theta-t) + \frac{r^2}{2}] + [\frac{1}{2} - \frac{r^2}{2}]}{1 - 2r\cos(\theta - t) + r^2} dt$$

$$= \int_{\alpha}^{\beta} [\tfrac{1}{2} + \tfrac{1}{2}\mathcal{P}(r, \theta - t)]dt = \pi m(S) + \tfrac{1}{2} \int_{\alpha}^{\beta} \mathcal{P}(r, \theta - t)dt,$$

we see that

$$m(S') = \frac{1}{2\pi} \int_{\alpha}^{\beta} \mathcal{P}(r, \theta - t)dt.$$

Therefore $\int_A \mathcal{P}(r, \theta - t)dt \geq \frac{1}{2\pi} \int_{\alpha}^{\beta} \mathcal{P}(r, \theta - t)dt = m(S') \geq \frac{\sigma}{\pi}$ proving (2).

Let $\xi(\omega) = \inf\{t \geq 0 : \gamma_t(\omega) \in \partial B \backslash \Gamma\}$ be the first hitting time of $\gamma_t(\omega)$ on $\partial B \backslash \Gamma$, with the understanding that $\xi(\omega) = \infty$ in case γ_t does not hit $\partial B \backslash \Gamma$ before it hits Γ (so that $\gamma_{\tau(\omega)}(\omega) \in A'$). Because of the continuity of the Brownian paths, almost surely $\gamma_t(\omega)$ cannot impact the unit circle in a point of A unless it crosses $\partial B \backslash \Gamma$. We seek, then, to make the following calculation, which we will immediately justify and explain by means of the strong Markov property of Brownian motion:

$$m(A) = P\{\omega : \gamma_{\tau(\omega)}(\omega) \in A\}$$

$$= \int_{\partial B \setminus \Gamma} P\{\omega : \gamma_{z,\tau_z(\omega)}(\omega) \in A\} P\{\omega : \gamma_{\xi(\omega)}(\omega) \in dz\}$$

$$\geq \frac{\sigma}{\pi} \int_{\partial B \setminus \Gamma} P\{\omega : \gamma_{\xi(\omega)}(\omega) \in dz\}$$

$$= \frac{\sigma}{\pi} P\{\omega : \gamma_t(\omega) \in \partial B \setminus \Gamma \text{ for some } t \geq 0\}$$

$$\geq \frac{\sigma}{\pi} P\{\omega : u^*(\omega) > \lambda\}.$$

If we let $\Omega_\xi = \{\omega : \xi(\omega) < \infty\}$ and define $Y : \Omega_\xi \to \partial B \setminus \Gamma$ by $Y(\omega) = \gamma_{\xi(\omega)}(\omega)$ (so that Y gives the first point of impact, if any, of γ_t with $\partial B \setminus \Gamma$ before Γ), then Y determines a probability measure μ_Y on $\partial B \setminus \Gamma$ according to

$$\mu_Y(E) = P\{\omega \in \Omega_\xi : Y(\omega) \in E\} = P(Y^{-1}E).$$

The integrals over $\partial B \setminus \Gamma$ are to be understood as taken with respect to this measure μ_Y; thus

$$\int_{\partial B \setminus \Gamma} P\{\omega : \gamma_{z,\tau_z(\omega)}(\omega) \in A\} P\{\omega : \gamma_{\xi(\omega)}(\omega) \in dz\} = \int_{\partial B \setminus \Gamma} P\{\omega : \gamma_{z,\tau_z(\omega)}(\omega) \in A\} \, d\mu_Y(z),$$

and

$$\int_{\partial B \setminus \Gamma} P\{\omega : \gamma_{\xi(\omega)}(\omega) \in dz\} = \int_{\partial B \setminus \Gamma} d\mu_Y(z) = \mu_Y(\partial B \setminus \Gamma)$$

$$= P(\Omega_\xi) = P\{\omega : \gamma_t(\omega) \in \partial B \setminus \Gamma \text{ for some } t \geq 0\}.$$

This observation should make sense of the fourth step in the preceding computation. Since the first, third, and fifth steps are clear because of Corollary 3.6.2, (2), and (1), respectively, in order to complete the proof it remains only to show that

$$(3) \qquad P\{\omega : \gamma_{\tau(\omega)}(\omega) \in A\} = \int_{\partial B \setminus \Gamma} P\{\omega : \gamma_{z,\tau_z(\omega)}(\omega) \in A\} \, d\mu_Y(z).$$

Let us consider again the process

$$\tilde{\gamma}_{z,t}(\omega) = \gamma_{z,t\wedge\tau_z(\omega)}(\omega)$$

which has Γ as an absorbing barrier. Because the $\tilde{\gamma}_{z,t}$ still possess the conditional-probability form of the strong Markov property, as in 3.6 for $t \geq 0$ we have that

$$P\{\omega : \xi(\omega) < \infty \text{ and } \tilde{\gamma}_{\xi(\omega)+t}(\omega) \in A\} = \int_{\Omega_\xi} P\{\tilde{\gamma}_{\xi+t} \in A \mid B_{\xi^+}\} dP$$

$$= \int_{\Omega_\xi} P_{Y(\omega)}\{\tilde{\gamma}_t \in A\} dP(\omega) = \int_{\Omega_\xi} P\{\omega' : \tilde{\gamma}_{Y(\omega),t}(\omega') \in A\} dP(\omega).$$

If we let t tend to infinity, the left side tends to $P\{\omega : \gamma_{\tau(\omega)}(\omega) \in A\}$, while the right side approaches

$$\int_{\Omega_\xi} P\{\omega' : \gamma_{Y(\omega),\tau_{Y(\omega)}(\omega')}(\omega') \in A\} dP(\omega)$$

$$= \int_{\partial B \setminus \Gamma} P\{\omega : \gamma_{z,\tau_z(\omega)}(\omega) \in A\} d\mu_Y(z),$$

and this proves (3).

4.2 Conditioned Brownian motion

We turn now to the proof of the left-hand inequality of Theorem 1, $c_\sigma m(A) \leq P\{\omega : u^*(\omega) > \lambda\}$, continuing with the notation established in 4.1. It is convenient also to continue working with the Brownian motions $\tilde{\gamma}_{z,t}$ which have Γ for an absorbing barrier, because after all we are only interested in what takes place in D. Denote by $X_z(\omega)$ the point of Γ at which $\tilde{\gamma}_{z,t}(\omega)$ is absorbed, so that

$$X_z(\omega) = \gamma_{z,\tau_z(\omega)}(\omega) = \tilde{\gamma}_{z,\infty}(\omega);$$

X_0 will be denoted simply by X.

For fixed $e^{i\theta} \in \Gamma$, P^θ will be the probability measure associated with the Brownian motion which starts at 0 and is <u>conditioned to hit</u> Γ <u>at</u> $e^{i\theta}$. Let us be more precise about this. The random variable X determines a measure μ_X on Γ by

$$\mu_X(F) = P\{\omega : X(\omega) \in F\}$$

for measurable $F \subset \Gamma$; because of Corollary 3.6.2, $\mu_X(F) = m(F)$. For fixed $E \in \mathcal{B}$ we may define a measure ν_E on Γ by

$$\nu_E(F) = \int_{X^{-1}F} P(E|X)dP = \int_{X^{-1}F} \chi_E dP = P(E \cap X^{-1}F).$$

Since ν_E is absolutely continuous with respect to $\mu_X = m$, by the Radon-Nikodym Theorem there is a measurable function $g_E : \Gamma \to R$ such that

$$\nu_E(F) = \int_F g_E(e^{i\theta})dm(\theta)$$

for all measurable $F \subset \Gamma$. Changing variables in the latter integral according to $e^{i\theta} = X(\omega)$ yields the equation

$$\int_{X^{-1}F} P(E|X)dP = \int_{X^{-1}F} g_E(X(\omega))dP(\omega),$$

and this implies that $g_E(X(\omega)) = P(E|X)(\omega)$ a. e. dP. Therefore, if we define

$$P^\theta(E) = g_E(e^{i\theta}),$$

it makes sense to think of $P^\theta(E)$ as the probability of the event E given that $X = e^{i\theta}$.

Fix $\theta \in [0, 2\pi)$, and recall that $\sigma \in (0, 1)$ is fixed. For each r with $\max\{\sigma, \frac{1}{2}\} \le r < 1$ let $S_r = \Omega_\sigma(\theta) \cap \{z : |z| = r\}$.

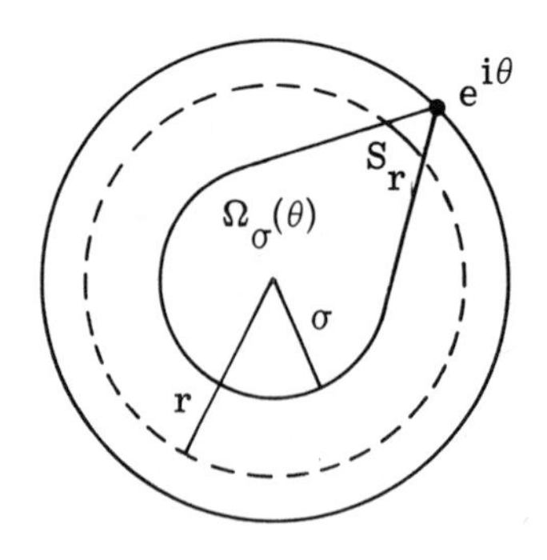

The rest of this section is devoted to proving that if σ is sufficiently small, then there is a constant $c_\sigma > 0$ such that

$$\text{(1)} \qquad P^\theta\{\omega : \tilde{\gamma}_t(\omega) \notin S_r \text{ for } t \geq 0\} \geq c_\sigma \text{ for all } r \text{ with } \max\{\sigma, \tfrac{1}{2}\} \leq r < 1.$$

In the following section the restrictions that σ be close to 0 and r be no less than $\frac{1}{2}$ will be removed.

For each r with $\max\{\sigma, \frac{1}{2}\} \leq r < 1$ let

$$E_r = \{\omega : \tilde{\gamma}_t(\omega) \in S_r \text{ for some } t \geq 0\}.$$

If we can show that for small σ there is a constant $b_\sigma < 1$ such that

(2) for all r, $P(E_r \cap X^{-1}F) \leq b_\sigma m(F)$ for every short arc $F \subset \Gamma$, then it will follow that

$$\int_F [P^\theta(E_r) - b_\sigma] dm(\theta) \leq 0 \text{ for short arcs } F,$$

and hence $P^\theta(E_r) \leq b_\sigma$ a. e. on Γ. Therefore, since $P^\theta(E_r)+P^\theta(\Omega \backslash E_r)=1$ a. e., (1) is an immediate consequence of (2).

Fix r and let $z \in S_r$. It was noted in the proof of 4.1 (2) that if F is an arc on Γ and F' is its projection through z, then

$$P\{\omega : X_z(\omega) \in F\} = m(F').$$

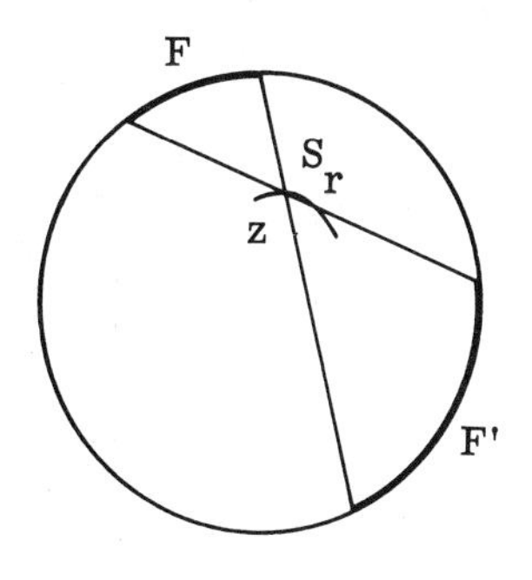

Clearly among all arcs F which have the same measure, the maximum of $m(F')$ is attained when F is centered at $\arg z$. Then similar triangles show that

$$m(F') \le 4\frac{1+r}{1-r}m(F)$$

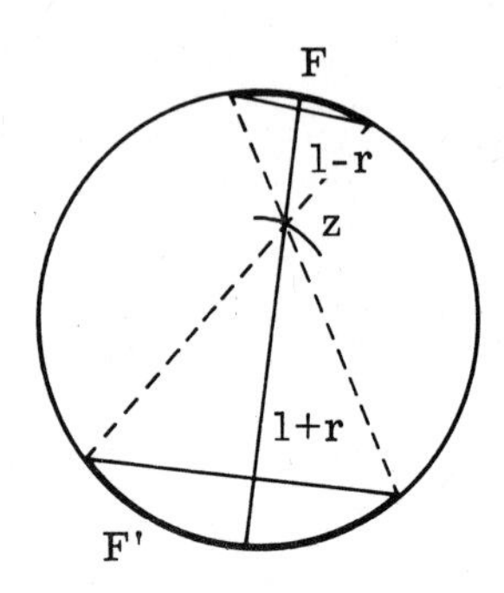

for every sufficiently short arc F (where the concept of 'sufficiently short' may depend on r).

Let $Y(\omega)$ be the first point of impact, if any, of $\tilde{\gamma}_t(\omega)$ on S_r. If $\mu_Y(G) = P(Y^{-1}G)$ for $G \subset S_r$, then as before we may use the strong Markov property to find that

$$(3) \qquad P(E_r \cap X^{-1}F) = \int_{S_r} P\{\omega : X_z(\omega) \in F\}\, d\mu_Y(z) \le 4\frac{1+r}{1-r}m(F)\mu_Y(S_r)$$
$$= 4\frac{1+r}{1-r}m(F)P(E_r)$$

for every short arc F.

Further progress towards the proof of (2) depends, then, upon estimating $P(E_r)$, which is the probability that γ_t hits S_r before it hits Γ. Let $\phi(z)$ be the harmonic function on $D\backslash S_r$ which has boundary values 1 on S_r and 0 on Γ. By Theorem 3.6.1, $P(E_r) = \phi(0)$. Because of the rotational invariance of Brownian motion, we may assume for the moment that $\theta = 0$, so that S_r is centered at the point r on the real axis.

The map

$$w = f(z) = \frac{z - r}{1 - rz}$$

is one-to-one conformal from D to D with $f(r) = 0$ and $f(1) = 1$. Under the inverse map

$$z = \frac{r + w}{1 + rw}$$

the image of a circle of radius ε centered at the origin is a circle C_ε containing the point r.

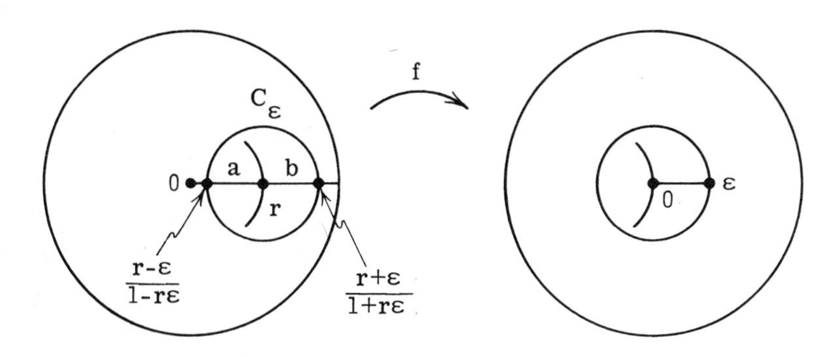

The distances a and b shown on the figure are given by

$$a = \varepsilon \frac{1 - r^2}{1 - r\varepsilon} \quad \text{and} \quad b = \varepsilon \frac{1 - r^2}{1 + r\varepsilon} \quad .$$

Let us choose ε so that C_ε is just large enough to contain S_r in its interior. Geometrical considerations will show that 0 is not inside C_ε, but this is clearly true for small σ, since $r \geq \frac{1}{2}$.

Let $\psi(w)$ be the harmonic function on $\{w : \varepsilon < |w| < 1\}$ which assumes boundary values 1 on $\{w : |w| = \varepsilon\}$ and 0 on $\{w : |w| = 1\}$. By either the probabilistic interpretation or known facts about harmonic functions, $\phi(z) \leq \psi(f(z))$ where both are defined. Since it is known that

$$\psi(w) = \frac{\log |w|}{\log \varepsilon} ,$$

it follows that

$$\phi(z) \leq \frac{\log \left| \frac{z - r}{1 - rz} \right|}{\log \varepsilon}$$

and

$$(4) \qquad P(E_r) = \phi(0) \leq \frac{\log r}{\log \varepsilon} .$$

Now one half the length of S_r must be no less than the smaller of a and b, which is b, so

$$l(S) \geq \varepsilon(1 - r^2).$$

Similar triangles show that

$$\frac{l(S)}{1 - r} \leq c(2\sigma),$$

so that certainly

$$l(S) \leq 8\sigma(1 - r)$$

for small σ. Combining these estimates with (4) shows that

$$\phi(0) \leq \frac{\log r}{\log(8\sigma) - \log(1 + r)} \leq c \frac{\log r}{\log \sigma}$$

for some constant c, so long as σ is close to 0, and hence, by (3),

$$P(E_r \cap X^{-1}F) \leq c \frac{1 + r}{1 - r} \frac{\log r}{\log \sigma} m(F).$$

Here c is an absolute constant and σ is required to be small. Now $\log r/(1 - r)$ is bounded for $\frac{1}{2} \leq r < 1$, and $\log \sigma \rightarrow -\infty$ as $\sigma \rightarrow 0$, so we may choose σ_0 so that

$$\left| c \frac{1 + r}{1 - r} \frac{\log r}{\log \sigma} \right| \leq b_{\sigma_0} < 1 \text{ whenever } \tfrac{1}{2} \leq r < 1 \text{ and } 0 < \sigma \leq \sigma_0.$$

This observation completes the proof of (2).

4.3 Removal of the restrictions on σ and r

In most arguments involving the regions $\Omega_\sigma(\theta)$, the choice of $\sigma \in (0, 1)$ is of minimal importance, affecting only moderately the constants associated with the various estimates. It might be supposed that the requirement in the preceding section that $r \geq \frac{1}{2}$ is more substantial, but it too can easily be erased. The reason for this is, of course, that it is the values that functions assume near Γ that are of importance, while their behavior at points far inside D is of less interest. For any σ and a with $0 < \sigma < 1$ and $\sigma \leq a < 1$, it is really enough to consider the maximal function

$$N_{\sigma,a}f(e^{i\theta}) = \sup\{|f(z)| : |z| \geq a, \ z \in \Omega_\sigma(\theta)\}$$

of a function f defined on D.

Lemma 4.3.1. For each fixed σ and a as above there is a constant $c_{\sigma,a}$ such that whenever u is harmonic on D and $\lambda > 0$,

$$m\{e^{i\theta} : N_\sigma u(e^{i\theta}) > \lambda\} \leq c_{\sigma,a} m\{e^{i\theta} : N_{\sigma,a}u(e^{i\theta}) > \lambda\}.$$

Proof. We may choose n and $\alpha_1, \ldots, \alpha_n \in [0, 2\pi)$ so that if

$$R_{\sigma,a}(\theta) = \bigcup_{k=1}^{n} [e^{i\alpha_k}\Omega_\sigma(\theta)] \text{ for } 0 \leq \theta \leq 2\pi,$$

then for all θ

$$\partial R_{\sigma,a}(\theta) \subset \{z : |z| \geq a\}.$$

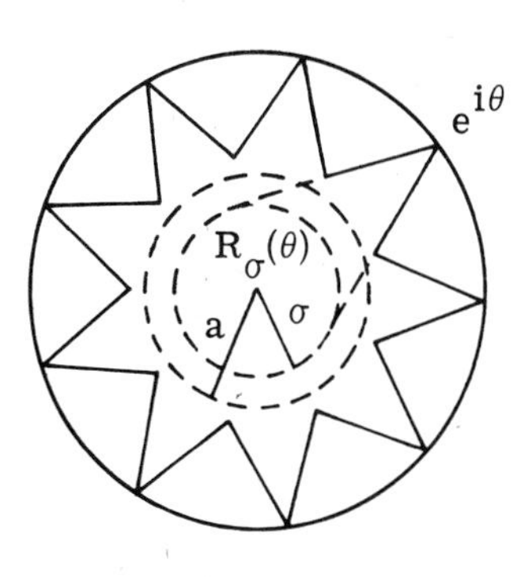

Then the Maximum Principle implies that

$$\{e^{i\theta} : N_\sigma u(e^{i\theta}) > \lambda\} \subset \{e^{i\theta} : \sup_{R_{\sigma,a}(\theta)} |u(z)| > \lambda\}$$

$$\subset \bigcup_{k=1}^{n} e^{-i\alpha_k}\{e^{i\theta} : N_{\sigma,a}u(e^{i\theta}) > \lambda\},$$

and hence

$$m\{e^{i\theta} : N_\sigma u(e^{i\theta}) > \lambda\} \leq nm\{e^{i\theta} : N_{\sigma,a} u(e^{i\theta}) > \lambda\}.$$

Because of this Lemma, in order to prove that $c_\sigma m(A) \leq P\{\omega : u^*(\omega) > \lambda\}$ it is enough to replace N_σ by $N_{\sigma,\frac{1}{2}}$ and prove instead that $c_\sigma m\{e^{i\theta} : N_{\sigma,\frac{1}{2}} u(e^{i\theta}) > \lambda\} \leq P\{\omega : u^*(\omega) > \lambda\}$; thereby the restriction that r be greater than or equal to $\frac{1}{2}$ is removed. A similar argument will dispense with the requirement that σ be small, once the following result has been established.

Lemma 4.3.2. <u>For each σ and σ_0 with $0 < \sigma_0 \leq \sigma < 1$ there is a constant c_{σ,σ_0} such that whenever u is harmonic on D and</u> $\lambda > 0$,

$$m\{e^{i\theta} : N_\sigma u(e^{i\theta}) > \lambda\} \leq c_{\sigma,\sigma_0} m\{e^{i\theta} : N_{\sigma_0} u(e^{i\theta}) > \lambda\}.$$

Proof. For each $z \in D$ with $|z| > \sigma$ let

$$T_\sigma z = \{e^{i\theta} : z \in \Omega_\sigma(\theta)\},$$

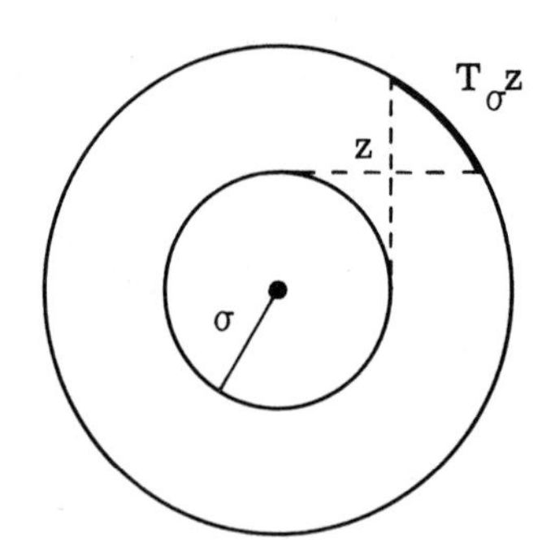

so that $T_\sigma z$ is an arc on Γ, centered at $\arg z$, of measure

$$\lambda_{\sigma,r} = \frac{1}{\pi}[\sin^{-1}\frac{\sigma}{r} - \sin^{-1}(r \sin(\pi - \sin^{-1}\frac{\sigma}{r}))],$$

where $r = |z|$. For fixed σ, $\lambda_{\sigma,r}$ is a decreasing function of r with limit 0 as r tends to 1, and an application of L'Hôpital's Rule shows that

$$\sup_{\sigma<r<1} \frac{\lambda_{\sigma, r}}{\lambda_{\sigma_0, r}} = c_{\sigma, \sigma_0} < \infty.$$

Lemma 4.3.1 implies that it is enough to prove that

$$m\{e^{i\theta} : N_{\sigma, \frac{1+\sigma}{2}} u(e^{i\theta}) > \lambda\} \le c_{\sigma, \sigma_0} m\{e^{i\theta} : N_{\sigma_0, \frac{1+\sigma}{2}} u(e^{i\theta}) > \lambda\}.$$

Now

$$\{e^{i\theta} : N_{\sigma, \frac{1+\sigma}{2}} u(e^{i\theta}) > \lambda\} \subset \cup\{T_\sigma z : |z| \ge \frac{1+\sigma}{2},\ |u(z)| > \lambda\}$$

by the Maximum Principle. Let $\mathfrak{D}$ be the operator which replaces each subarc of Γ by one c_{σ, σ_0} times as long but with the same center. We know, then, that

$$\mathfrak{D}(T_{\sigma_0} z) \supset T_\sigma z \text{ for all } z \text{ with } \sigma < |z| < 1.$$

If $\{S_\alpha\}$ is a family of disjoint subarcs of Γ such that

$$\{e^{i\theta} : N_{\sigma_0, \frac{1+\sigma}{2}} u(e^{i\theta}) > \lambda\} = \bigcup_\alpha S_\alpha,$$

then, because $\mathfrak{D}(S_\alpha) \supset \cup\{\mathfrak{D}(T_{\sigma_0} z) : T_{\sigma_0} z \subset S_\alpha\}$ for each α, it follows that

$$m\{e^{i\theta} : N_{\sigma, \frac{1+\sigma}{2}} u(e^{i\theta}) > \lambda\} \le m(\cup\{T_\sigma z : |z| \ge \frac{1+\sigma}{2},\ |u(z)| > \lambda\})$$

$$\le m(\cup\{\mathfrak{D}(T_{\sigma_0} z) : |z| \ge \frac{1+\sigma}{2},\ |u(z)| > \lambda\})$$

$$\le m(\bigcup_\alpha \mathfrak{D}(S_\alpha))$$

$$\le \sum_\alpha m(\mathfrak{D}(S_\alpha)) \le c_{\sigma, \sigma_0} \sum_\alpha m(S_\alpha)$$

$$= c_{\sigma, \sigma_0} m\{e^{i\theta} : N_{\sigma_0, \frac{1+\sigma}{2}} u(e^{i\theta}) > \lambda\}.$$

Theorem 4.1 will be proved, then, if we can show that $c_\sigma m\{e^{i\theta} : N_{\sigma, \frac{1}{2}} u(e^{i\theta}) > \lambda\} \le P\{\omega : u^*(\omega) > \lambda\}$ whenever σ is close to 0.

4.4 Conditionally reflected Brownian motion

For each $t \geq 0$, let $\hat{\gamma}_t(\omega)$ be the reflection of $\tilde{\gamma}_t(\omega)$ about the segment from 0 to $X(\omega)$ (the point of impact of $\tilde{\gamma}_t(\omega)$ on Γ). If $\tilde{\gamma}_t(\omega) = r_t(\omega)e^{i\theta_t(\omega)}$ and $X(\omega) = e^{i\theta(\omega)}$, then

$$\hat{\gamma}_t(\omega) = r_t(\omega)e^{i(2\theta(\omega)-\theta_t(\omega))} = X^2(\omega)\overline{\gamma_t(\omega)}.$$

I assert that for each $\lambda > 0$,

$$(1) \qquad P^\theta\{\omega : \sup_{0 \leq t < \infty} |u(\tilde{\gamma}_t(\omega))| > \lambda\}$$
$$= P^\theta\{\omega : \sup_{0 \leq t < \infty} |u(\hat{\gamma}_t(\omega))| > \lambda\} \text{ a.e. } dm(\theta).$$

Let us suppose for a moment that this result has been established. If σ is in the range prescribed in 4.2 and $N_{\sigma,\frac{1}{2}}u(e^{i\theta}) > \lambda$, then there is an $r \in [\frac{1}{2}, 1)$ such that $S_r = \Omega_\sigma(\theta) \cap \{z : |z| = r\}$ contains a point z for which $|u(z)| > \lambda$. If $\{\tilde{\gamma}_t(\omega) : t \geq 0\}$ does not hit S_r, then $\{\tilde{\gamma}_t(\omega) : t \geq 0\} \cup \{\hat{\gamma}_t(\omega) : t \geq 0\}$ contains a closed curve surrounding S_r on which, by the Maximum Principle, u must assume values larger than λ. Let $^*u(\omega) = \sup\{|u(\hat{\gamma}_t(\omega))| : t \geq 0\}$.

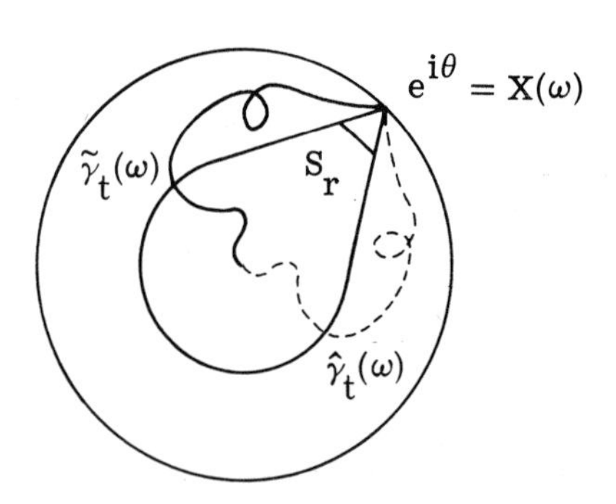

Then for almost all such θ, by (1) and 4.2 (1),

$$P^\theta\{\omega : u^*(\omega) > \lambda\} = \tfrac{1}{2}(P^\theta\{\omega : u^*(\omega) > \lambda\} + P^\theta\{\omega : {}^*u(\omega) > \lambda\})$$
$$\geq \tfrac{1}{2}P^\theta\{\omega : u^*(\omega) \vee {}^*u(\omega) > \lambda\}$$
$$\geq \tfrac{1}{2}P^\theta\{\omega : \gamma_t(\omega) \notin S_r \text{ for } t \geq 0\} \geq c_\sigma > 0.$$

If $K = \{e^{i\theta} : N_{\sigma,\frac{1}{2}}u(e^{i\theta}) > \lambda\}$, then

$$P\{\omega : u^*(\omega) > \lambda\} = \int_\Omega P(u^* > \lambda | X)dP$$
$$= \int_\Gamma P^\theta\{\omega : u^*(\omega) > \lambda\}dm(\theta) \geq \int_K P^\theta\{\omega : u^*(\omega) > \lambda\}\, dm(\theta)$$
$$\geq c_\sigma m(K).$$

We have reduced the proof of Theorem 4. 1, then, to that of statement (1).

[This calculation works because P^θ disregards all those paths for which $X(\omega) \neq e^{i\theta}$ - since P^θ is defined by

$$P(E \cap X^{-1}F) = \int_F P^\theta(E)dm(\theta) \quad \text{for } E \subset \Omega \text{ and } F \subset \Gamma,$$

only those paths which land in F influence the values of P^θ on F. More precisely, we will mention below that once certain sets of measure 0 are discarded, each P^θ becomes a measure on Ω. Since for measurable $F, G \subset \Gamma$,

$$\int_G P^\theta(X^{-1}F)dm(\theta) = P(X^{-1}F \cap X^{-1}G) = m(F \cap G),$$

$P^\theta(X^{-1}F) = \chi_F(e^{i\theta})$ a. e. Thus if $\{F_n\}$ is a decreasing sequence of arcs with $\cap F_n = \{e^{i\theta}\}$, then

$$P^\theta(X^{-1}e^{i\theta}) = P^\theta(\cap X^{-1}F_n) = \lim P^\theta(X^{-1}F_n) = 1.$$

When applying P^θ to any event, then, we are correct in ignoring all paths ω except those for which $X(\omega) = e^{i\theta}$.]

Now even more than (1) is true, in that $\{\hat{\gamma}_t : t \geq 0\}$ is again Brownian motion absorbed at its first point of impact on Γ. Let us consider the unabsorbed reflected process

$$\gamma_t^*(\omega) = X^2(\omega)\,\overline{\gamma_t(\omega)} \qquad (t \geq 0);$$

we wish to prove that for any Borel set $U \subset C$ and $s, t \geq 0$,

$$(2) \qquad P\{\omega : \gamma^*_{s+t}(\omega) - \gamma^*_t(\omega) \in U\} = P\{\omega : \gamma_{s+t}(\omega) - \gamma_t(\omega) \in U\}.$$

Since clearly $\gamma^*_0 = 0$ a.e., and a similar argument (with U replaced by a Borel subset of C^k) to the following one shows that $\{\gamma^*_t\}$ has independent increments, (2) will imply that $\{\gamma^*_t\}$ is again two-dimensional Brownian motion.

The argument can be completed easily once we believe that it is permissible to treat X as a constant when we condition on it; for Brownian motion is invariant under conjugation and rotation so for each fixed $\alpha \in R$,

$$P(\gamma_{s+t} - \gamma_t \in e^{i\alpha}\overline{U}|X) = P(\gamma_{s+t} - \gamma_t \in U|X) \quad \text{a. e.}$$

and hence

$$P(\gamma_{s+t} - \gamma_t \in X^2\,\overline{U}|X) = P(\gamma_{s+t} - \gamma_t \in U|X) \quad \text{a. e.}$$

Once this has been established, it will follow easily that

$$\begin{aligned} P\{\omega : \gamma^*_{s+t}(\omega) - \gamma^*_t(\omega) \in U\} &= \int P(\gamma^*_{s+t} - \gamma^*_t \in U|X)dP \\ &= \int P(\gamma_{s+t} - \gamma_t \in X^2\,\overline{U}|X)dP \\ &= \int P(\gamma_{s+t} - \gamma_t \in U|X)dP = P\{\omega : \gamma_{s+t}(\omega) - \gamma_t(\omega) \in U\}. \end{aligned}$$

It is not easy, however, to prove that a random variable on which one is conditioning can indeed be treated as momentarily constant, since some subtle topological and measure-theoretic questions must be confronted first.

Recall that P^θ is defined by

$$P(E \cap X^{-1}F) = \int_F P^\theta(E)dm(\theta)$$

for measurable $E \subset \Omega$ and $F \subset \Gamma$. This implies that

$$P^\theta(E) \geq 0 \quad \text{a. e.} \quad dm(\theta) \quad \text{and}$$

$$P^\theta(\cup E_i) = \sum P^\theta(E_i) \quad \text{a. e.} \quad dm(\theta) \quad \text{for disjoint measurable } E_1, E_2, \ldots .$$

Now our underlying probability space $(\Omega, \mathcal{B}, P)$ as we constructed it is the infinite product of Borel spaces (a Borel space is a Borel subset of a complete metric space with the σ-algebra of its Borel subsets), and hence $(\Omega, \mathcal{B})$ is a Borel space. Further, by the continuity of the sample paths, the process $\{\gamma_t\}$ is separable, in that $\mathcal{B}$ is generated by all sets of the form

$$\{\omega : \gamma_{s_i+t_j}(\omega) - \gamma_{t_i}(\omega) \in A_k\},$$

where the A_k are chosen from a countable open base for C and s_i and t_j from a countable dense set in $[0, \infty)$. In any case, this situation is sufficiently regular that it is possible to discard sets of measure 0 in such a way that P^θ is a measure on $(\Omega, \mathcal{B})$ for each θ [40, §§26 and 27], [48, V. 8]. Assume that this has been done.

For each Borel set $A \subset C$ and $\omega \in \Omega$ let

$$Q(\omega, A) = P^{X(\omega)}\{\omega' : \gamma_{t+s}(\omega') - \gamma_t(\omega') \in A\}.$$

Then for fixed ω, $Q(\omega, A)$ is a Borel measure on C; and since for fixed $\alpha \in R$, $e^{2i\alpha}\bar{\gamma}_t$ is again Brownian motion on C,

$$Q(\omega, e^{2i\alpha}\bar{A}) = Q(\omega, A) \text{ for all } \omega \text{ and } A.$$

Now for measurable $F \subset \Gamma$ and $U \subset C$,

$$\begin{aligned}
&P\{\omega : X(\omega) \in F \text{ and } \gamma_{t+s}(\omega) - \gamma_t(\omega) \in U\} \\
&= \int_F P^\theta\{\omega : \gamma_{t+s}(\omega) - \gamma_t(\omega) \in U\}dm(\theta) \\
&= \int_{X^{-1}F} P^{X(\omega)}\{\omega' : \gamma_{t+s}(\omega') - \gamma_t(\omega') \in U\}dP(\omega) = \int_{X^{-1}F} Q(\omega,U)dP(\omega) \\
&= \int_{X^{-1}F}\int_U Q(\omega, dz)dP(\omega) = \int_{X^{-1}F}\int_C \chi_U(z)Q(\omega, dz)dP(\omega) .
\end{aligned}$$

This implies that if

$$F(\theta, \omega) = \int_C \chi_U(z)Q(\omega, dz)$$

(which of course is independent of θ), then

$$P(\gamma_{t+s} - \gamma_t \in U | X)(\omega) = F(\theta, \omega) \text{ a. e. } dP(\omega) .$$

Note also that because of the invariance property of Q, for each $\alpha \in R$

$$F(\theta, \omega) = \int_C \chi_U(e^{2i\alpha}\bar{z})Q(\omega, dz) ,$$

and in particular if

$$G(e^{i\theta}, \omega) = \int_C \chi_U(e^{2i\theta}\bar{z})Q(\omega, dz) ,$$

then $G(e^{i\theta}, \omega) = F(\theta, \omega)$.

Define $Q^*(\omega, A)$, for $\omega \in \Omega$ and Borel $A \subset C$, by requiring that

$$P\{\omega : X(\omega) \in F \text{ and } \gamma_{s+t}(\omega) - \gamma_t(\omega) \in X^2(\omega)\bar{A}\} = \int_{X^{-1}F} Q^*(\omega,A)dP(\omega)$$

for each measurable $F \subset \Gamma$. Again for almost all fixed ω, $Q^*(\omega, A)$ is a measure on C; moreover, the idempotent change of variables $z \to X^2(\omega)\bar{z}$ interchanges $Q(\omega, \cdot)$ and $Q^*(\omega, \cdot)$. Therefore, for $F \subset \Gamma$,

$$P\{\omega : X(\omega) \in F \text{ and } \gamma^*_{t+s}(\omega) - \gamma^*_t(\omega) \in U\}$$

$$= \int_{X^{-1}F} Q^*(\omega, U)dP(\omega) = \int_{X^{-1}F} \int_C \chi_U(z)Q^*(\omega, dz)dP(\omega)$$

$$= \int_{X^{-1}F} \int_C \chi_U(X^2(\omega)\bar{z})Q(\omega, dz)dP(\omega) = \int_{X^{-1}F} G(X(\omega), \omega)dP(\omega) ,$$

implying that

$$P(\gamma^*_{t+s} - \gamma^*_t \in U | X)(\omega) = G(X(\omega), \omega) \text{ a. e. } dP(\omega) .$$

We conclude that

$$P(\gamma^*_{t+s} - \gamma^*_t \in U | X) = P(\gamma_{t+s} - \gamma_t \in U | X) \text{ a. e.}$$

and

$$P\{\omega : \gamma^*_{t+s}(\omega) - \gamma^*_t(\omega) \in U\} = \int P(\gamma^*_{t+s} - \gamma^*_t \in U | X)dP$$

$$= \int P(\gamma_{t+s} - \gamma_t \in U | X)dP = P\{\omega : \gamma_{t+s}(\omega) - \gamma_t(\omega) \in U\},$$

proving (2).

5·Inequalities for the conjugate function

The next step in the proof of Theorem 6.1 is to show that for $0 < p < \infty$ there is no great difference between the L^p norms of the Brownian maximal functions of a harmonic function and its conjugate. The argument, carried out mainly on Ω, uses stopping times, the conformal invariance of the Brownian paths, and a result (Lemma 5.1.4) reminiscent of some standard martingale inequalities (one of which appears in 5.2). In Section 2, with the help of the maximal inequality for martingales we obtain as a corollary M. Riesz' inequalities for the conjugate function (2.5). Once the connection between the norms of the maximal functions of conjugates has been established, the proof of Theorem 6.1 can be completed (in the next chapter) with no great difficulty.

5.1 A Brownian maximal function inequality

The proof of the following result will be based on three lemmas and several changes of variables.

Theorem 5.1.1. <u>For each</u> p <u>with</u> $0 < p < \infty$ <u>there is a constant</u> c_p <u>such that whenever</u> u <u>is harmonic on</u> D <u>and</u> $\tilde{u}$ <u>is its harmonic conjugate with</u> $\tilde{u}(0) = 0$,

$$\|\tilde{u}^*\|_p \leq c_p \|u^*\|_p .$$

Let $F = u + i\tilde{u}$ be the analytic completion of the given harmonic function u. There is no loss of generality in assuming that $F(0) = 0$. It will be convenient to restrict our attention to closed disks slightly smaller than D.

Fix r with $0 < r < 1$ and let $\tau_r(\omega) = \inf\{t \geq 0 : |\gamma_t(\omega)| \geq r\}$ be the first hitting time of $\gamma_t(\omega)$ on $\Gamma_r = \{z : |z| = r\}$. Let

$\alpha_t(\omega) = \gamma_{t\wedge\tau_r(\omega)}(\omega)$ be the Brownian motion starting at 0 with absorbing barrier Γ_r, and let

$$F_r^*(\omega) = \sup\{|F(\alpha_t(\omega))| : t \geq 0\} \text{ and}$$

$$u_r^*(\omega) = \sup\{|u(\alpha_t(\omega))| : t \geq 0\}$$

be the maximal functions, with respect to $\{\alpha_t\}$, of F and u, respectively. We will prove that for each p there is a constant c_p independent of r and u such that $\|F_r^*\|_p \leq c_p \|u^*\|_p$; since $\tilde{u}^* \leq F^*$, this is enough to prove Theorem 5.1.1.

Lemma 5.1.2. Suppose that ξ and η are stopping times of $\{\alpha_t\}$ with $\xi(\omega) \leq \eta(\omega)$ a.e. Then

$$\|\tilde{u} \circ \alpha_\eta - \tilde{u} \circ \alpha_\xi\|_2 = \|u \circ \alpha_\eta - u \circ \alpha_\xi\|_2 .$$

Proof. By the conformal invariance of the Brownian paths (3.3.6), $\{F^2(\gamma_t) : t \geq 0\}$ is again Brownian motion, until time τ and up to an order-preserving change of time scale. Therefore (for further details see Ch. 7) $\{F^2(\alpha_t) : t \geq 0\}$ is a martingale, as are $\{F^2(\alpha_{t\wedge\xi}) : t \geq 0\}$ and $\{F^2(\alpha_{t\wedge\eta}) : t \geq 0\}$ [30, p. 71], [14, p. 302]. (Note that the random variables are integrable because they are bounded.) In particular, for each $t > 0$

$$\int F^2(\alpha_{t\wedge\xi(\omega)}(\omega))dP(\omega) = \int F^2(\alpha_{0\wedge\xi(\omega)}(\omega)dP(\omega) = 0,$$

and

$$0 = \lim_{t\to\infty} \int F^2(\alpha_{t\wedge\xi(\omega)}(\omega))dP(\omega) = \int \lim_{t\to\infty} F^2(\alpha_{t\wedge\xi(\omega)}(\omega))dP(\omega)$$

$$= \int F^2(\alpha_{\xi(\omega)}(\omega))dP(\omega) = \int [u^2(\alpha_\xi) - \tilde{u}^2(\alpha_\xi) + 2iu(\alpha_\xi)\tilde{u}(\alpha_\xi)]dP,$$

so that $\|u \circ \alpha_\xi\|_2 = \|\tilde{u} \circ \alpha_\xi\|_2$. A similar calculation shows that $\|u \circ \alpha_\eta\|_2 = \|\tilde{u} \circ \alpha_\eta\|_2$.

Let us consider now a stochastic process $\{x_0, x_1, x_2\}$ on Ω which consists of just three functions: $x_0(\omega) = 0$, $x_1(\omega) = u(\alpha_{\xi(\omega)}(\omega))$, and $x_2(\omega) = u(\alpha_{\eta(\omega)}(\omega))$. Since $\{F(\gamma_t) : t \geq 0\}$ is two-dimensional

Brownian motion with a new clock, its real part $\{u(\gamma_t) : t \geq 0\}$ must be one-dimensional Brownian motion with a new clock and hence is still a martingale. Then [20, p. 193], [14, p. 302] the process $\{x_0, x_1, x_2\}$ is again a martingale. In particular,

$$E(u(\alpha_\eta) \mid u(\alpha_\xi)) = u(\alpha_\xi),$$

so that

$$u(\alpha_\xi)^2 = u(\alpha_\xi)E(u(\alpha_\eta) \mid u(\alpha_\xi)) = E(u(\alpha_\xi)u(\alpha_\eta) \mid u(\alpha_\xi)),$$

and integrating over Ω shows that

$$E(u(\alpha_\xi)u(\alpha_\eta)) = E(u(\alpha_\xi)^2) .$$

The same argument proves also that

$$E(\tilde{u}(\alpha_\xi)\tilde{u}(\alpha_\eta)) = E(\tilde{u}(\alpha_\xi)^2) .$$

Therefore

$$\|u(\alpha_\eta) - u(\alpha_\xi)\|_2^2 = \int [u(\alpha_\eta)^2 - 2u(\alpha_\eta)u(\alpha_\xi) + u(\alpha_\xi)^2]dP$$

$$= \int [u(\alpha_\eta)^2 - u(\alpha_\xi)^2]dP = \int [\tilde{u}(\alpha_\eta)^2 - \tilde{u}(\alpha_\xi)^2]dP$$

$$= \|\tilde{u}(\alpha_\eta) - \tilde{u}(\alpha_\xi)\|_2^2.$$

We include the easy proof of the following result of Paley and Zygmund [59, V, 8.26] for the sake of completeness.

Lemma 5.1.3. <u>Let</u> g <u>be a measurable function on a measure space</u> $(X, \mathcal{B}, \mu)$ <u>and</u> $E \subset X$ <u>a measurable set with</u> $\mu(E) > 0$ <u>and</u> $g \geq 0$ <u>on</u> E. <u>Suppose further that</u> A <u>and</u> B <u>are constants such that</u>

(i) $\quad \frac{1}{\mu(E)} \int_E g \geq A > 0$ <u>and</u>

(ii) $\quad \frac{1}{\mu(E)} \int_E g^2 \leq B.$

<u>Then for any</u> δ <u>with</u> $0 < \delta < 1$,

$$\mu\{x : g(x) \geq \delta A\} \geq \mu(E)(1 - \delta)^2 \frac{A^2}{B} .$$

Proof. If $E_\delta = \{x \in E : g(x) \geq \delta A\}$, then

$$\int_{E \setminus E_\delta} g \leq \int_{E \setminus E_\delta} \delta A \leq \delta A \mu(E),$$

so by (i)

$$A\mu(E) \leq \int_E g = \int_{E \setminus E_\delta} g + \int_{E_\delta} g \leq \delta A \mu(E) + \int_{E_\delta} g,$$

and hence

$$\int_{E_\delta} g \geq A\mu(E)(1 - \delta) .$$

But the Schwarz Inequality and (ii) imply that

$$\int_{E_\delta} g \leq (\int_{E_\delta} g^2)^{\frac{1}{2}} (\mu(E_\delta))^{\frac{1}{2}} \leq (B\mu(E))^{\frac{1}{2}} (\mu(E_\delta))^{\frac{1}{2}}.$$

Therefore

$$A\mu(E)(1 - \delta) \leq (B\mu(E))^{\frac{1}{2}} (\mu(E_\delta))^{\frac{1}{2}}$$

and

$$\mu(E_\delta) \geq \frac{A^2}{B} (1 - \delta)^2 \mu(E),$$

as required.

Lemma 5.1.4. _Let_ $a \geq 1$ _and_ $b > 1$. _There are constants_ $c_{a,b}$ _and_ $d_{a,b}$ _such that if_ $\lambda > 0$ _and_

$$P\{\omega : F_r^*(\omega) > \lambda\} \leq aP\{\omega : F_r^*(\omega) > b\lambda\},$$

then

$$P\{\omega : F_r^*(\omega) > \lambda\} \leq c_{a,b} P\{\omega : u_r^*(\omega) > d_{a,b}\lambda\}.$$

Proof. We define stopping times ξ and η of $\{\alpha_t\}$ by

$$\xi(\omega) = \inf\{t \geq 0 : |F(\alpha_t(\omega))| > \lambda\}$$

and

$$\eta(\omega) = \inf\{t \geq 0 : |F(\alpha_t(\omega))| > b\lambda\},$$

so that $\xi \le \eta$ a.e. (Note that ξ and η may be $+\infty$ for some paths ω.) On $\{\omega : F_r^* > b\lambda\} = \{\omega : \eta(\omega) < \infty\}$ we have $|F(\alpha_\xi)| = \lambda$ and $|F(\alpha_\eta)| = b\lambda$. Also, if $F_r^*(\omega) \le \lambda$, then $\xi(\omega) = \eta(\omega) = \infty$ and $\alpha_{\xi(\omega)}(\omega) = \alpha_{\eta(\omega)}(\omega) = \gamma_{\tau_r(\omega)}(\omega)$. Therefore, by Lemma 5.1.2,

$$\int_{\{F_r^* > \lambda\}} [u(\alpha_\eta) - u(\alpha_\xi)]^2 dP = \|u(\alpha_\eta) - u(\alpha_\xi)\|_2^2$$

$$= \tfrac{1}{2}\|F(\alpha_\eta) - F(\alpha_\xi)\|_2^2 \ge \tfrac{1}{2}\int_{\{F_r^* > b\lambda\}} |F(\alpha_\eta) - F(\alpha_\xi)|^2 dP$$

$$\ge \tfrac{1}{2}(b\lambda - \lambda)^2 P\{F_r^* > b\lambda\} \ge \tfrac{1}{2}\frac{(b-1)^2}{a}\lambda^2 P\{F_r^* > \lambda\},$$

by hypothesis. Also,

$$\int_{\{F_r^* > \lambda\}} [u(\alpha_\eta) - u(\alpha_\xi)]^4 dP \le \int_{\{F_r^* > \lambda\}} |F(\alpha_\eta) - F(\alpha_\xi)|^4 dP$$

$$= \int_{\{F_r^* > b\lambda\}} |F(\alpha_\eta) - F(\alpha_\xi)|^4 dP + \int_{\{\lambda < F_r^* \le b\lambda\}} |F(\alpha_\eta) - F(\alpha_\xi)|^4 dP$$

$$\le (b\lambda + \lambda)^4 P\{F_r^* > b\lambda\} + 16 b^4 \lambda^4 P\{\lambda < F_r^* \le b\lambda\}$$

$$\le [(b+1)^4 + 16 b^4]\lambda^4\, P\{F_r^* > \lambda\}.$$

Now according to Lemma 5.1.3 with

$$g(\omega) = [u(\alpha_{\eta(\omega)}(\omega)) - u(\alpha_{\xi(\omega)}(\omega))]^2, \quad E = \{\omega : F_r^*(\omega) > \lambda\},$$

$$A = \tfrac{1}{2}\frac{(b-1)^2}{a}\lambda^2, \quad \text{and} \quad B = [(b+1)^4 + 16 b^4]\lambda^4,$$

we may conclude that for any δ with $0 < \delta < 1$,

$$P\{F_r^* > \lambda\} \le \frac{B}{A^2(1-\delta)^2} P\left\{|u(\alpha_\eta) - u(\alpha_\xi)| \ge \left|\frac{\delta}{2}\frac{(b-1)^2}{a}\right|^{\frac{1}{2}}\lambda\right\}.$$

If we fix a value of δ and note that $|u(\alpha_\eta) - u(\alpha_\xi)| \le 2u_r^*$, the conclusion follows immediately.

Proof of Theorem 5.1.1. Fix p with $0 < p < \infty$. If for $x \ge 0$

$$G(x) = P\{F_r^* > x\},$$

then

$$\|F_r^*\|_p^p = \int_\Omega |F_r^*|^p dP = -\int_0^\infty \lambda^p dG(\lambda) ,$$

and integration by parts, remembering that F is bounded on $\{z : |z| \le r\}$ so that $G(x) = 0$ for large x, converts this last integral to

$$\int_0^\infty G(\lambda)p\lambda^{p-1}d\lambda = \int_0^\infty p\lambda^{p-1}P\{F_r^* > \lambda\}d\lambda.$$

Now let $a = 2^{p+1}$ and $b = 2$. If B denotes the set of all $\lambda > 0$ for which

$$P\{F_r^* > \lambda\} \le aP\{F_r^* > b\lambda\}$$

then

$$\int_0^\infty p\lambda^{p-1}P\{F_r^* > \lambda\}d\lambda = \int_0^\infty p(bx)^{p-1}P\{F_r^* > bx\}bdx$$

$$= b^p[\int_B px^{p-1}P\{F_r^* > bx\}dx + \int_{[0,\infty)\backslash B} px^{p-1}P\{F_r^* > bx\}dx]$$

$$\le b^p[\int_B px^{p-1}P\{F_r^* > x\}dx + \frac{1}{a}\int_{[0,\infty)\backslash B} px^{p-1}P\{F_r^* > x\}dx]$$

$$= b^p\int_B px^{p-1}P\{F_r^* > x\}dx + \frac{b^p}{a}\int_0^\infty px^{p-1}P\{F_r^* > x\}dx - \frac{b^p}{a}\int_B px^{p-1}P\{F_r^* > x\}dx$$

$$= b^p(1 - \frac{1}{a})\int_B px^{p-1}P\{F_r^* > x\}dx + \frac{b^p}{a}\int_0^\infty px^{p-1}P\{F_r^* > x\}dx,$$

so that

$$(1 - \frac{b^p}{a})\int_0^\infty px^{p-1}P\{F_r^* > x\}dx \le b^p(1 - \frac{1}{a})\int_B px^{p-1}P\{F_r^* > x\}dx$$

and

$$\int_0^\infty px^{p-1}P\{F_r^* > x\}dx \le \frac{b^p(1 - \frac{1}{a})}{(1 - \frac{b^p}{a})}\int_B px^{p-1}P\{F_r^* > x\}dx$$

$$\le a\int_B px^{p-1}P\{F_r^* > x\}dx .$$

Thus we may apply Lemma 5.1.4 to see that

$$\|F_r^*\|_p^p = \int_0^\infty p\lambda^{p-1}P\{F_r^* > \lambda\}d\lambda \le a\int_B px^{p-1}P\{F_r^* > x\}dx$$

$$\le a\int_B px^{p-1}c_{a,b}P\{u_r^* > d_{a,b}x\}dx$$

$$\le \frac{ac_{a,b}}{d_{a,b}}\int_0^\infty px^{p-1}P\{u_r^* > x\}dx = c_p\|u_r^*\|_p^p \le c_p\|u^*\|_p^p.$$

Note that we have actually proved that $\|\tilde{u}_r^*\|_p \le c_p \|u_r^*\|_p$ for $0 \le r < 1$; this observation will be used in the next section.

5.2 The M. Riesz inequalities

In order to obtain Riesz' inequalities for the conjugate function (2.5) as a corollary of Theorem 5.1.1, we have to appeal to Doob's maximal inequality for submartingales. For the convenience of the reader we include the proof, which is based on the following easy maximal inequality.

Lemma 5.2.1. _Let_ $\{x_1, x_2, \ldots, x_n\}$ _be a submartingale,_ $\lambda \in \mathbf{R}$, _and_ $E_\lambda = \{\omega : \sup_{1\le k\le n} x_k(\omega) \ge \lambda\}$. _Then_

$$\lambda P(E_\lambda) \le \int_{E_\lambda} x_n dP .$$

Proof. For $k = 1, 2, \ldots, n$ let

$$A_k = \{\omega : x_k(\omega) \ge \lambda \text{ but } x_j(\omega) < \lambda \text{ if } j < k\}.$$

Then $A_k \in \mathcal{B}(x_1, \ldots, x_k)$ and E_λ is the disjoint union of $A_1, \ldots, A_n$. Therefore

$$\int_{E_\lambda} x_n dP = \sum_{k=1}^{n} \int_{A_k} x_n dP = \sum_{k=1}^{n} \int_{A_k} E(x_n | x_1, \ldots, x_k) dP$$
$$\ge \sum_{k=1}^{n} \int_{A_k} x_k dP \ge \lambda \sum_{k=1}^{n} \int_{A_k} dP = \lambda P(E_\lambda) .$$

This Lemma extends with a slight change to the case of a submartingale $\{x_1, x_2, \ldots\}$ with infinitely many nonnegative terms. For given $\lambda \in \mathbf{R}$, let $E_{n,\lambda} = \{\omega : \sup_{1\le k\le n} x_k(\omega) \ge \lambda\}$. Then

$$\lambda P(E_{n,\lambda}) \le \int_{E_{n,\lambda}} (\sup_{1\le k\le n} x_k) dP$$

by the Lemma, and since $E_{1,\lambda} \subset E_{2,\lambda} \subset \ldots$ the Monotone Convergence Theorem implies that

$$\text{(1)} \qquad \lambda P\{\sup_{k\ge 1} x_k \ge \lambda\} \le \int_{\{\sup_{k\ge1} x_k \ge \lambda\}} (\sup_{k\ge 1} x_k) dP .$$

Theorem 5.2.2, The Maximal Inequality for Submartingales [14, p. 317]. Let $\{x_t : a \leq t \leq b\}$ be a submartingale such that, for almost all ω, $x_t(\omega)$ is a right-continuous function of t, and suppose that $x_t \geq 0$ a.e. for all t. Then for $p > 1$,

$$\| \sup_t x_t \|_p \leq \frac{p}{p-1} \| x_b \|_p .$$

Proof. Let $\{t_1, t_2, \ldots, t_n\}$ be a finite subset of $[a, b]$ with $t_1 < t_2 < \ldots < t_n$, and let $y_k = x_{t_k}$ for $k = 1, 2, \ldots, n$. We will show first that

$$\| \sup_{1 \leq k \leq n} y_k \|_p \leq \frac{p}{p-1} \| y_n \|_p ,$$

and a density argument will complete the proof.

Let $Y = \sup_{1 \leq k \leq n} y_k$. If $G(\lambda) = P\{Y > \lambda\}$, then as in the proof of Theorem 5.1.1

$$E(Y^p) = -\int_0^\infty \lambda^p dG(\lambda) = \int_0^\infty p\lambda^{p-1} G(\lambda) d\lambda - \lambda^p G(\lambda) \Big|_0^\infty$$

$$\leq \int_0^\infty p\lambda^{p-1} G(\lambda) d\lambda = \int_0^\infty p\lambda^{p-1} P\{Y > \lambda\} d\lambda .$$

By the Lemma, this last integral is no larger than

$$\int_0^\infty p\lambda^{p-1} \frac{1}{\lambda} \int_{\{Y \geq \lambda\}} y_n dP d\lambda = p \int_\Omega y_n(\omega) \int_0^{Y(\omega)} \lambda^{p-2} d\lambda dP(\omega)$$

$$= \frac{p}{p-1} \int_\Omega y_n(\omega) Y(\omega)^{p-1} dP(\omega) \leq \frac{p}{p-1} \| y_n \|_p \| Y^{p-1} \|_q$$

$$= q \| y_n \|_p \, E(Y^p)^{1/q},$$

where the last inequality is accomplished by Hölder's Inequality, $q = \frac{p}{p-1}$, and the obvious convention applies in case some integrals should turn out to be $+\infty$. Therefore

$$E(Y^p)^{1-1/q} \leq q \| y_n \|_p,$$

and the first step in the proof is complete.

We have shown that for each finite subset T of $[a, b]$ which includes b,

$$\|\sup_{t\in T} x_t\|_p \leq \frac{p}{p-1} \|x_b\|_p .$$

Let D be a countable dense subset of [a, b]. Since D is the increasing union of finite sets T_n which contain b, for each ω we have

$$\sup_{t\in D} x_t(\omega) = \lim_{n\to\infty} \sup_{t\in T_n} x_t(\omega) ,$$

and so by the Monotone Convergence Theorem

$$\|\sup_{t\in D} x_t\|_p = \lim_{n\to\infty} \|\sup_{t\in T_n} x_t\|_p \leq \frac{p}{p-1} \|x_b\|_p .$$

But for all ω not in a certain set of measure 0,

$$\sup_{t\in D} x_t(\omega) = \sup_{a\leq t\leq b} x_t(\omega) ,$$

and this completes the proof.

Theorem 5. 2. 3, M. Riesz' Inequalities. <u>For each</u> p <u>with</u> $1 < p < \infty$ <u>there is a constant</u> c_p <u>such that whenever</u> u <u>is harmonic on</u> D, $\tilde{u}$ <u>is its harmonic conjugate with</u> $\tilde{u}(0) = 0$, <u>and</u> $0 \leq r < 1$,

$$\frac{1}{2\pi} \int_0^{2\pi} |\tilde{u}(re^{i\theta})|^p d\theta \leq c_p \frac{1}{2\pi} \int_0^{2\pi} |u(re^{i\theta})|^p d\theta .$$

Proof. Fix $r \in [0, 1)$ and as in 5. 1 let $\tau_r(\omega) = \inf\{t\geq 0 : |\gamma_t(\omega)| \geq r\}$ be the first hitting time of $\gamma_t(\omega)$ on $\Gamma_r = \{z : |z| = r\}$, and continue to let $\alpha_t = \gamma_{t\wedge\tau_r}$ be the Brownian motion which starts at the origin and is absorbed at its first point of impact on Γ_r. By projection and conformal invariance, $\{u(\alpha_t) : t \geq 0\}$ is a martingale (see the proof of Lemma 5. 1. 2), and hence $\{|u(\alpha_t)|^p : t \geq 0\}$ is a submartingale [14, p. 296] (see also p. 70). For each $n \geq 0$, because of the Maximal Inequality for Submartingales

$$\| \sup_{0\leq t\leq n} |u(\alpha_t)| \|_p \leq \frac{p}{p-1} \|u(\alpha_n)\|_p ;$$

hence, by the Bounded Convergence Theorem,

$$\| \sup_{0\leq t<\infty} |u(\alpha_t)| \|_p \leq \frac{p}{p-1} \|u(\alpha_{\tau_r})\|_p .$$

Now

$$[M_p(r, \tilde{u})]^p = \frac{1}{2\pi} \int_0^{2\pi} |\tilde{u}(re^{i\theta})|^p d\theta = r \int_\Omega |\tilde{u}(\alpha_{\tau_r})|^p dP$$

$$= r \|\tilde{u}(\alpha_{\tau_r})\|_p^p \le r \| \sup_{0 \le t < \infty} |\tilde{u}(\alpha_t)| \|_p^p$$

$$\le rc_p \| \sup_{0 \le t < \infty} |u(\alpha_t)| \|_p^p ,$$

by Theorem 5.1.1. Because of our foregoing remarks, this is no greater than

$$c_p \left(\frac{p}{p-1}\right)^p r \|u(\alpha_{\tau_r})\|_p^p = c_p \left(\frac{p}{p-1}\right)^p [M_p(r, u)]^p ,$$

and the proof is complete.

6·The maximal function characterization of H^p

Now we are in a position to give the proof of the Burkholder-Gundy-Silverstein Theorem. Only in the proof of the old (left-hand) inequality do we need to appeal to a fact from probability theory that has not yet been encountered, and the proof of this result is provided below.

Theorem 6.1. For each p with $0 < p < \infty$ and σ with $0 < \sigma < 1$ there are constants $c_{\sigma,p}$ and $C_{\sigma,p}$ such that whenever u is harmonic on D, $\tilde{u}$ is its harmonic conjugate with $\tilde{u}(0) = 0$, and $F = u + i\tilde{u}$ is its analytic completion,

$$c_{\sigma,p} \int_0^{2\pi} [N_\sigma u(e^{i\theta})]^p d\theta \leq \sup_{0 \leq r < 1} \int_0^{2\pi} |F(re^{i\theta})|^p d\theta \leq C_{\sigma,p} \int_0^{2\pi} [N_\sigma u(e^{i\theta})]^p d\theta.$$

Proof. We will consider the right-hand inequality first. As before, for $\lambda \in R$ let $G(\lambda) = P\{u^* > \lambda\}$; then

$$\|u^*\|_p^p = \int_\Omega (u^*)^p dP = -\int_0^\infty \lambda^p dG(\lambda)$$

$$= \int_0^\infty p\lambda^{p-1} G(\lambda) d\lambda - \lambda^p G(\lambda)\Big|_0^\infty \leq \int_0^\infty p\lambda^{p-1} G(\lambda) d\lambda$$

$$= \int_0^\infty p\lambda^{p-1} P\{u^* > \lambda\} d\lambda,$$

and by Theorem 4.1 this is less than or equal to

$$c_\sigma \int_0^\infty p\lambda^{p-1} m\{e^{i\theta} : N_\sigma u(e^{i\theta}) > \lambda\} d\lambda .$$

Now if $\int_0^{2\pi} [N_\sigma u]^p d\theta = +\infty$, there is nothing to prove, so we assume that $N_\sigma u \in L^p(\Gamma)$; hence, if $H(\lambda) = m\{e^{i\theta} : N_\sigma u(e^{i\theta}) > \lambda\}$, then

$$\int_0^{2\pi} [N_\sigma u(e^{i\theta})]^p dm(\theta) = -\int_0^\infty \lambda^p dH(\lambda) < \infty .$$

Consequently,

$$0 = \lim_{\lambda_0 \to \infty} -\int_{\lambda_0}^{\infty} \lambda^p dH(\lambda) \geq \lim_{\lambda_0 \to \infty} \lambda_0^p(-\int_{\lambda_0}^{\infty} dH(\lambda))$$

$$= \lim_{\lambda_0 \to \infty} \lambda_0^p H(\lambda_0),$$

and we see that

$$\|u^*\|_p^p \leq c_\sigma \int_0^\infty p\lambda^{p-1} m\{e^{i\theta} : N_\sigma u(e^{i\theta}) > \lambda\} d\lambda$$

$$= c_\sigma \int_0^\infty p\lambda^{p-1} H(\lambda) d\lambda = c_\sigma [\lambda^p H(\lambda)\Big|_0^\infty - \int_0^\infty \lambda^p dH(\lambda)]$$

$$= c_\sigma \int_0^{2\pi} [N_\sigma u(e^{i\theta})]^p dm(\theta) = \frac{c_\sigma}{2\pi} \int_0^{2\pi} [N_\sigma u(e^{i\theta})]^p d\theta \ .$$

Since by Theorem 5.1.1 $\|\tilde{u}^*\|_p^p \leq c_p \|u^*\|_p^p$, it follows that also

$$\|\tilde{u}^*\|_p^p \leq c_{\sigma, p} \int_0^{2\pi} [N_\sigma u(e^{i\theta})]^p d\theta \ .$$

Now if $0 \leq r < 1$ and $\tau_r(\omega) = \inf\{t \geq 0 : |\gamma_t(\omega)| \geq r\}$, then

$$\frac{1}{2\pi} \int_0^{2\pi} |\tilde{u}(re^{i\theta})|^p d\theta = r\|\tilde{u}(\gamma_{\tau_r})\|_p^p \leq \|\tilde{u}^*\|_p^p,$$

and similarly

$$\frac{1}{2\pi} \int_0^{2\pi} |u(re^{i\theta})|^p d\theta \leq \|u^*\|_p^p \ .$$

Therefore

$$\frac{1}{2\pi} \int_0^{2\pi} |F(re^{i\theta})|^p d\theta = \frac{1}{2\pi} \int_0^{2\pi} |u(re^{i\theta}) + i\tilde{u}(re^{i\theta})|^p d\theta$$

$$\leq \frac{1}{2\pi} \int_0^{2\pi} 2^p (\max\{|u(re^{i\theta})|, |\tilde{u}(re^{i\theta})|\})^p d\theta$$

$$\leq \frac{1}{2\pi} \int_0^{2\pi} 2^p (|u(re^{i\theta})|^p + |\tilde{u}(re^{i\theta})|^p) d\theta$$

$$\leq C_{\sigma, p} \int_0^{2\pi} [N_\sigma u(e^{i\theta})]^p d\theta \ .$$

Turning now to the proof of the left-hand inequality (the half of this theorem that was already known to Hardy and Littlewood), suppose that $F \in H^p$. It follows from a result of Doob, which we will examine in more detail below, that if again $\alpha_t = \gamma_{t \wedge \tau_r}$, then $\{|F(\alpha_t)|^{p/2} : t \geq 0\}$ is a submartingale. Therefore the Maximal Inequality implies that for

$0 \le r < 1$,

$$\| \sup_{0 \le t < \tau_r} |u(\gamma_t)| \|_p \le \| \sup_{0 \le t < \tau_r} |F(\gamma_t)|^{p/2} \|_2^{2/p}$$

$$\le c_p \| |F(\gamma_{\tau_r})|^{p/2} \|_2^{2/p} = c_p [\frac{1}{2\pi r} \int_0^{2\pi} |F(re^{i\theta})|^p d\theta]^{1/p}$$

(where the second inequality is established as in the proof of Riesz' Inequalities in 5.2); and because of the monotone convergence as $r \to 1$,

$$\|u^*\|_p \le c_p \lim_{r \to 1} [\frac{1}{2\pi r} \int_0^{2\pi} |F(re^{i\theta})|^p d\theta]^{1/p}$$

$$= c_p [\sup_{0 \le r < 1} \frac{1}{2\pi} \int_0^{2\pi} |F(re^{i\theta})|^p d\theta]^{1/p} .$$

In particular, $\|u^*\|_p < \infty$, so the proof may be completed by the argument that worked successfully on the first half: if $G(\lambda) = P\{u^* > \lambda\}$, then $\lim_{\lambda \to \infty} \lambda^p G(\lambda) = 0$ and

$$\|u^*\|_p^p = - \int_0^\infty \lambda^p dG(\lambda) = \int_0^\infty p\lambda^{p-1} G(\lambda) d\lambda - \lambda^p G(\lambda) \Big|_0^\infty$$

$$= \int_0^\infty p\lambda^{p-1} P\{u^* > \lambda\} d\lambda \ge c_{\sigma, p} \int_0^\infty p\lambda^{p-1} m\{e^{i\theta} : N_\sigma u(e^{i\theta}) > \lambda\} d\lambda$$

$$\ge c_{\sigma, p} \| N_\sigma u(e^{i\theta}) \|_p^p .$$

Recall that a continuous function ϕ defined on an open connected set $U \subset C$ is said to be subharmonic in case for each $z \in U$ there is an R such that $\{\zeta : |\zeta - z| \le R\} \subset U$ and

$$\phi(z) \le \frac{1}{2\pi} \int_0^{2\pi} \phi(z + re^{i\theta}) d\theta \quad \text{for all } r \text{ with } 0 < r \le R.$$

A function ϕ is subharmonic on U if and only if each harmonic function that dominates it on the boundary of a subdomain dominates it on the interior as well. If F is analytic on U and $c > 0$, then $|F|^c$ is subharmonic on U. The following result will complete the proof of Theorem 6.1.

Theorem 6.2 [15, Th. 4.3]. If ϕ is subharmonic on $D_r = \{z : |z| < r\}$ and continuous on $\overline{D}_r$, then $\{\phi(\alpha_t) : t \ge 0\}$ is a submartingale.

Proof. Let $0 \leq s_1 < s_2 < \ldots < s_n < t$, and suppose that $G \in \mathcal{B}(\phi(\alpha_{s_1}), \ldots, \phi(\alpha_{s_n}))$. We will show that

$$\int_G \phi(\alpha_t(\omega))dP(\omega) \geq \int_G \phi(\alpha_{s_n}(\omega))dP(\omega) ;$$

this is enough to imply that $E(\phi(\alpha_t) | \phi(\alpha_{s_1}), \ldots, \phi(\alpha_{s_n})) \geq \phi(\alpha_{s_n})$ a. e.

Let $\tau_{z,r} = \inf\{t \geq 0 : |\gamma_{z,t}| \geq r\}$ be the first hitting time on Γ_r of the Brownian motion starting at z, $\alpha_{z,t} = \gamma_{z, t \wedge \tau_{z,r}}$, and $Y(\omega) = \alpha_{s_n}(\omega)$. There is a density function $\delta(t, \rho)$ such that for $A \subset D$ and $t \geq 0$,

$$P\{\omega : \alpha_t(\omega) \in A\} = \iint_A \delta(t, \rho)\rho d\rho d\psi .$$

As in 3. 5. 4, by the strong Markov property of $\{\alpha_t : t \geq 0\}$, if $\mathcal{B}_n = \mathcal{B}\{\phi(\alpha_s) : 0 \leq s \leq s_n\}$, then

$$\int_G \phi(\alpha_t)dP = \int_G E(\phi(\alpha_t) | \mathcal{B}_n)dP = \int_G E(\phi(\alpha_{Y(\omega), t-s_n}))dP$$

$$= \int_G [\int_\Omega \phi(\alpha_{Y(\omega), t-s_n}(\omega'))dP(\omega')]dP(\omega).$$

For fixed ω, let $\Omega_1 = \{\omega' : t - s_n < \tau_{Y(\omega)}(\omega')\}$ and $\Omega_2 = \{\omega' : t - s_n \geq \tau_{Y(\omega)}(\omega')\}$. Using 3. 5. 4 once more, for $T \geq 0$ and $z \in D_r$

$$\int_{\Omega_2} \phi(\alpha_{z, T+(t-s_n)})dP = \int_{\Omega_2} E(\phi(\alpha_{z, T+(t-s_n)}) | \alpha_s, 0 \leq s \leq s_n)dP$$

$$= \int_{\Omega_2} E(\phi(\alpha_{z,T}))dP = P(\Omega_2)E(\phi(\alpha_{z,T}));$$

letting T tend to infinity shows that

$$\int_{\Omega_2} \phi(\alpha_{z,\infty})dP = P(\Omega_2) \int_\Omega \phi(\alpha_{z,\infty})dP .$$

Since

$$\int_{\Omega_2} \phi(\alpha_{Y(\omega), t-s_n})dP = \int_{\Omega_2} \phi(\alpha_{Y(\omega), \infty})dP ,$$

Theorem 3. 6. 1 implies that

$$\int_G \phi(\alpha_t)dP = \int_G [\int_{\Omega_1} \phi(\alpha_{Y(\omega), t-s_n}(\omega') + \int_{\Omega_2} \phi(\alpha_{Y(\omega), t-s_n}(\omega'))dP(\omega')]dP(\omega)$$

$$= \int_G [\int_0^{r-|Y(\omega)|} \int_0^{2\pi} \phi(Y(\omega) + \rho e^{i\psi})\delta(t - s_n, \rho)\rho d\rho d\psi$$

$$+ P(\Omega_2) \int_\Omega \phi(\alpha_{Y(\omega), \infty}(\omega'))dP(\omega')]dP(\omega)$$

$$\geq \int_G [\phi(Y(\omega)) \int_0^{r-|Y(\omega)|} 2\pi\delta(t-s_n, \rho)\rho d\rho + P(\Omega_2)\frac{1}{2\pi}\int_0^{2\pi}\phi(\rho e^{i\psi})\wp(|Y(\omega)|, \theta-\psi)d\psi]dP,$$

since ϕ is subharmonic (where $\theta = \arg Y(\omega)$); and this is no less than

$$\int_G [\phi(Y(\omega))P(\Omega_1) + P(\Omega_2)\phi(Y(\omega))]dP = \int_G \phi(\alpha_{s_n}(\omega))dP(\omega).$$

7·The Martingale versions of H^p and BMO

The proof of Theorem 6.1 shows that for $0 < p < \infty$ an analytic function $F = u + i\tilde{u}$ is in H^p if and only if u^*, the Brownian maximal function of the real part of F, is in $L^p(\Omega)$. Furthermore, for $F \in H^p$ the norms $\|u^*\|_{L^p(\Omega)}$, $\|N_\sigma u\|_{L^p(\Gamma)}$, $\|F^*\|_{L^p(\Omega)}$, $\|F\|_{H^p}$, and $\|F(e^{i\theta})\|_{L^p(\Gamma)}$ are all equivalent. This observation permits the identification of H^p, for $p \geq 1$, with a certain space of martingales on $(\Omega, \mathcal{B}, P)$. By analogy, then, the concept of 'H^p martingale' can be carried over to the case of a general measure space. Similarly, there is a definition of bounded mean oscillation (BMO) martingales such that under the correct identifications the BMO martingales on $(\Omega, \mathcal{B}, P)$ correspond to the BMO functions on Γ. The Fefferman-Stein Theorem, which identifies BMO as the dual of H^1, holds true in the martingale setting as well; here we prove the analytic theorem by probabilistic methods.

In what follows, from time to time the same symbol is used to denote a harmonic (or analytic) function and its boundary function.

7.1 H^p martingales

Let $F = u + i\tilde{u} \in H^p$ for some $p \geq 1$, so that $F \in H^1$ as well, and for $t \geq 0$ let $\tilde{\mathcal{B}}_t = \mathcal{B}\{\tilde{\gamma}_s : 0 \leq s \leq t\}$. Let

$$x_t(\omega) = u(\tilde{\gamma}_t(\omega)) \text{ for } t \geq 0;$$

as in 5.1, $\{x_t : t \geq 0\}$ is a martingale with respect to $\{\tilde{\mathcal{B}}_t : t \geq 0\}$. Let us examine the reason for this in more detail. First notice that $x_t \in L^1$ for all $t \geq 0$, and in fact $\|x_t\|_{L^1(\Omega)}$ is a bounded function of t, since

$$\int_\Omega |x_t| dP = \int_\Omega |u(\tilde{\gamma}_t)| dP = \iint_D |u(z)| P\{\tilde{\gamma}_t \in dA(z)\} + \int_\Gamma |u(z)| P\{\gamma_\tau \in dz\}$$

$$\leq \iint_D |u(re^{i\theta})| \frac{1}{2\pi t} e^{-r^2/2t} r dr d\theta + \frac{1}{2\pi} \int_0^{2\pi} |u(e^{i\theta})| d\theta$$

$$\leq 2\|F\|_{H^1} \quad \text{for all } t.$$

The martingale property follows from Doob's optional sampling theorem [14, p. 365 ff.] as follows. There is a time-change function $T(t, \omega)$ defined on $\Omega_t = \{\omega : t < \tau(\omega)\}$ such that

$$F(\gamma_t(\omega)) = \gamma_{T(t, \omega)}(\omega) \quad \text{on } \Omega_t$$

(see 3. 3. 6). Since $T(t, \omega)$ is increasing in t for each ω and the Brownian paths are continuous,

$$\lim_{t \to \tau(\omega)^-} T(t, \omega) = T(\tau(\omega), \omega)$$

exists a. e. and

$$F(\gamma_{\tau(\omega)}(\omega)) = \gamma_{T(\tau(\omega), \omega)}(\omega) \quad \text{a. e.}$$

Then the process

$$z_s(\omega) = \gamma_{s \wedge T(\tau(\omega), \omega)}(\omega)$$

is a martingale, and so is its real part $x_t(\omega) = u(\tilde{\gamma}_t(\omega))$, where t corresponds to s under the order-preserving map $s = T(t, \omega)$. Because $F \in H^p$, the function

$$x^*(\omega) = \sup_{t \geq 0} |x_t(\omega)|$$

is in $L^p(\Omega)$. Note also that x_t converges a. e. to $x_\infty = \phi(\gamma_\tau)$, where ϕ is the boundary function of u, so that in particular x_∞ is measurable with respect to $\mathcal{B}(\gamma_\tau)$.

We show now that this situation can be reversed, in that the original H^p function is recoverable from the martingale which it generates. Suppose then that $\{x_t : t \geq 0\}$ is a real-valued martingale on $(\Omega, \mathcal{B}, P)$ with respect to $\{\tilde{\mathcal{B}}_t : t \geq 0\}$ and that $x^* = \sup_{t \geq 0} |x_t(\omega)| \in L^p(\Omega)$ for some

$p \geq 1$. Then

$$E(|x_t|) \leq [E(|x_t|^p)]^{1/p} \leq [E((x^*)^p)]^{1/p} < \infty \text{ for all } t \geq 0,$$

and the Submartingale Convergence Theorem (3.4.1) implies that $\{x_t\}$ converges a.e. to a function $x_\infty \in L^p(\Omega)$. Moreover, since

$$\int_{\{|x_t|>n\}} |x_t| dP \leq \int_{\{x^*>n\}} x^* dP \to 0,$$

$x_t \to x_\infty$ in $L^1(\Omega)$ [14, p. 319]. Let us assume further that x_∞ is measurable with respect to $\mathcal{B}(\gamma_\tau)$.

We claim now that for any function $f \in L^1(\Gamma)$ and $s, t \geq 0$,

$$(1) \qquad E(f(\tilde{\gamma}_{t+s}) | \tilde{\mathcal{B}}_t) = E(f(\tilde{\gamma}_{t+s}) | \tilde{\gamma}_t) \text{ a.e.}$$

For by Corollary 3.5.4,

$$E(f(\tilde{\gamma}_{t+s}) | \tilde{\mathcal{B}}_t)(\omega) = \int_\Omega f(\tilde{\gamma}_{\tilde{\gamma}_t(\omega), s}(\omega')) dP(\omega') \text{ a.e.};$$

since this is measurable with respect to $\mathcal{B}(\tilde{\gamma}_t) \subset \tilde{\mathcal{B}}_t$, (1) follows immediately. Note that if we let s tend to infinity, we obtain

$$(2) \qquad E(f(\gamma_\tau) | \tilde{\mathcal{B}}_t) = E(f(\gamma_\tau) | \tilde{\gamma}_t) \text{ a.e.}$$

Since x_∞ is measurable with respect to $\mathcal{B}(\gamma_\tau)$, there is a real-valued measurable function ϕ on Γ such that $x_\infty = \phi(\gamma_\tau)$ a.e. (This function is determined a.e. by the condition

$$\int_{\gamma_\tau^{-1} E} x_\infty dP = \int_E \phi dm$$

for all measurable $E \subset \Gamma$.) Of course $\phi \in L^p(\Gamma)$ since $x_\infty \in L^p(\Omega)$. If we define

$$u(re^{i\theta}) = \int_0^{2\pi} \phi(e^{i\psi}) \mathcal{P}(r, \theta - \psi) dm(\psi),$$

then u is harmonic on D. We claim that

$$x_t = u(\tilde{\gamma}_t) \text{ a.e.};$$

this will imply that $u^* \in L^p(\Omega)$ and hence $u + i\tilde{u} \in H^p$.

Since x_t and $E(x_\infty | \tilde{\mathcal{B}}_t)$ are both $\tilde{\mathcal{B}}_t$-measurable, and the martingale property implies that for each $A \in \tilde{\mathcal{B}}_t$

$$\int_A x_\infty dP = \lim_{s\to\infty} \int_A x_s dP = \lim_{s\to\infty} \int_A E(x_s | \tilde{\mathcal{B}}_t) dP$$

$$= \lim_{s\to\infty} \int_A x_t dP = \int_A x_t dP,$$

it must be the case that

$$x_t = E(x_\infty | \tilde{\mathcal{B}}_t) \text{ a. e.}$$

By (2), then, also

$$x_t = E(x_\infty | \tilde{\gamma}_t) \text{ a. e.}$$

Since $u(\tilde{\gamma}_t)$ is clearly measurable with respect to $\mathcal{B}(\tilde{\gamma}_t)$, in order to prove that it coincides a. e. with x_t it is enough to show that

$$\int_A u(\tilde{\gamma}_t) dP = \int_A x_\infty dP \text{ for all } A \in \mathcal{B}(\tilde{\gamma}_t).$$

Suppose then that $A_0 \subset C$ is measurable and $A = \{\omega : \gamma_t(\omega) \in A_0\}$. Let μ_t be the measure determined on C by $\tilde{\gamma}_t$ according to the equation

$$\mu_t(G) = P\{\omega : \tilde{\gamma}_t(\omega) \in G\}$$

for each measurable $G \subset C$. Then changes of variables allow us to write

$$\int_A u(\tilde{\gamma}_t) dP = \int_{A_0} u(z) d\mu_t(z) = \int_{A_0} \left[\frac{1}{2\pi} \int_0^{2\pi} \phi(e^{i\psi}) \mathcal{P}(r, \theta-\psi) d\psi\right] d\mu_t(z)$$

$$= \int_{A_0} \int_\Omega \phi(\tilde{\gamma}_{z,\tau}(\omega)) dP d\mu_t(z) = \int_A \int_\Omega \phi(\tilde{\gamma}_{\tilde{\gamma}_t(\omega'),\tau(\omega)}(\omega)) dP(\omega) dP(\omega').$$

Now Corollary 3.5.4 implies that for $s > 0$,

$$\int_A E(\phi(\tilde{\gamma}_{\tilde{\gamma}_t(\omega'),s})) dP(\omega') = \int_A E(\phi(\tilde{\gamma}_{s+t}) | \tilde{\mathcal{B}}_t) dP = \int_A \phi(\tilde{\gamma}_{s+t}) dP,$$

and letting s tend to infinity gives

$$\int_A E(\phi(\tilde{\gamma}_{\tilde{\gamma}_t(\omega'),\tau})) dP = \int_A \phi(\tilde{\gamma}_\tau) dP = \int_A x_\infty dP .$$

Thus

$$\int_A u(\tilde{\gamma}_t)dP = \int_A x_\infty dP,$$

so that $x_\infty = u(\tilde{\gamma}_t)$ a. e.

Since the norms $\|x^*\|_{L^p(\Omega)}$, $\|N_\sigma u\|_{L^p(\Gamma)}$, and $\|F\|_{H^p}$ are all equivalent, we have established the following result.

Theorem 7.1.1. *For* $p \geq 1$, H^p *is in one-to-one correspondence with the set of all real-valued martingales* $\{x_t : t \geq 0\}$ *on* $(\Omega, \mathcal{B}, P)$ *with respect to the family of* σ-*algebras* $\tilde{\mathcal{B}}_t = \mathcal{B}\{\tilde{\gamma}_s : 0 \leq s \leq t\}$ *for which*

$$x^* = \sup_{t \geq 0} |x_t| \in L^p(\Omega) \text{ and}$$

$$x_\infty = \lim_{t \to \infty} x_t \text{ is measurable with respect to } \mathcal{B}(\gamma_\tau).$$

The correspondence is effected by assigning to each $F \in H^p$ *the martingale*

$$x_t = \operatorname{Re} F(\tilde{\gamma}_t),$$

and to each martingale $\{x_t : t \geq 0\}$ *the analytic completion of*

$$u(re^{i\theta}) = \frac{1}{2\pi} \int_0^{2\pi} \phi(e^{i\psi})\mathcal{P}(r, \theta - \psi)d\psi, \text{ where } x_\infty = \phi(\gamma_\tau).$$

Moreover, the H^p *norm of* F *is equivalent to the* $L^p(\Omega)$ *norm of the Brownian maximal function of the corresponding martingale.*

The requirement that x_∞ be $\mathcal{B}(\gamma_\tau)$-measurable can be submerged by replacing x_t by $E(u(\tilde{\gamma}_t)|\mathcal{B}(\gamma_\tau))$ and $\mathcal{B}$ by $\mathcal{B}(\gamma_\tau)$. Then H^p corresponds to the set of martingales $\{x_t\}$ on $(\Omega, \mathcal{B}(\gamma_\tau), P)$ for which $x^* \in L^p(\Omega)$. This justifies the definition of H^p martingales in the general case.

Definition 7.1.2. Let $\{x_t : t \in T\}$ be a martingale on a probability space $(X, \mathcal{F}, \mu)$ with respect to an increasing family of σ-algebras $\{\mathcal{F}_t : t \in T\}$, and let $p \geq 1$. We say that $\{x_t : t \in T\}$ is *of class* H^p in case

$$x^* = \sup_{t \in T} |x_t| \in L^p(X).$$

7.2 BMO martingales

Definition 7.2.1. Let $\{x_t : t \in T\}$ be an H^2 martingale on a probability space $(X, \mathcal{F}, \mu)$ with respect to an increasing family of σ-algebras $\{\mathcal{F}_t : t \in T\}$, so that $x_t \to x_\infty$ for some $x_\infty \in L^2(X)$ and $x_t = E(x_\infty | \mathcal{F}_t)$ a.e. We say that $\{x_t : t \in T\}$ has <u>bounded mean oscillation</u>, and write $\{x_t\} \in \text{BMO}$, in case

$$(1) \qquad \sup_{t \in T} \| E(|x_\infty - x_t|^2 | \mathcal{F}_t) \|_{L^\infty(X)} < \infty .$$

This definition and most of the steps in the argument which supports it by linking it to the usual definition of a BMO function seem to be due to R. F. Gundy (unpublished). Recall that a function $f \in L^2(\Gamma)$ is said to have bounded mean oscillation $(f \in \text{BMO})$ in case

$$(2) \qquad \|f\|_{\text{BMO}} = \sup_I \frac{1}{m(I)} \int_I \left| f - \frac{1}{m(I)} \int_I f dm \right| dm < \infty ,$$

where the supremum is taken over all subintervals I of Γ. Thus a BMO function is one which on the average is not far from its average, near every point of Γ. The expressions appearing in (1) and (2) are not norms, being zero for constants, but can be converted to norms by adding $|E(x_\infty)|$ and $|\int_\Gamma f dm|$, respectively.

That these two versions of BMO are essentially the same is not at all obvious; but we will see, by means of several intermediate steps, that a (possibly complex-valued) martingale $\{f_t\}$ on $(\Omega, \mathcal{B}, P)$ with respect to $\{\tilde{\mathcal{B}}_t\}$ for which f_∞ is measurable with respect to $\mathcal{B}(\gamma_\tau)$, so that $f_\infty = g(\gamma_\tau)$ for some function g on Γ, satisfies (1) if and only if g satisfies (2). The key observation is that the interval averages

$$\frac{1}{m(I)} \int_I g dm = \int_\Gamma g \frac{\chi_I}{m(I)} dm$$

can be replaced by Poisson averages

$$\int_\Gamma g(e^{i\psi}) \mathcal{P}(r, \theta - \psi) dm(\psi) ,$$

so that $g \in \text{BMO}$ if and only if

$$(3) \qquad \sup_{r, \theta} \int_\Gamma \left| g(e^{i\sigma}) - \int_\Gamma g(e^{i\psi}) \mathcal{P}(r, \theta - \psi) dm(\psi) \right|^2 \mathcal{P}(r, \theta - \sigma) dm(\sigma) < \infty;$$

this relates the concept of bounded mean oscillation to a measure of the discrepancy of a harmonic function from its boundary values and, by introducing the Poisson kernel, makes possible changes of variables which carry us over to the probabilistic context. In general, any sufficiently sharply spiked approximate identity can be substituted for the Poisson kernel as well, without changing the class of functions so defined.

Some technical lemmas are needed to establish the different characterizations of bounded mean oscillation which concern us, which are stated as propositions and summarized in 7.3.3. The first important result on functions of bounded mean oscillation was the John-Nirenberg Theorem, which we recall here.

Theorem 7.2.2 **(John and Nirenberg [28]).** _A function f has bounded mean oscillation if and only if there are positive numbers b and N such that for each subinterval I of Γ_

$$m\{e^{i\sigma} \in I : |f(e^{i\sigma}) - \frac{1}{m(I)} \int_I f(e^{i\psi})dm(\psi)| > t\} \leq m(I)e^{-bt}$$

for all $t \geq N$.

Corollary 7.2.3. _Fix $p \geq 1$. A function f on Γ has bounded mean oscillation if and only if_

$$\sup_I \frac{1}{m(I)} \int_I |f(e^{i\sigma}) - \frac{1}{m(I)} \int_I f(e^{i\psi})dm(\psi)|^p dm(\sigma) < \infty.$$

Proof. That any function which satisfies this condition has bounded mean oscillation follows readily from Hölder's Inequality. Conversely, suppose that $f \in BMO$ and for each fixed $I \subset \Gamma$ let

$$\lambda_I(t) = m\{e^{i\sigma} \in I : |f(e^{i\sigma}) - \frac{1}{m(I)} \int_I f dm| > t\}.$$

By a change of variables and integration by parts,

$$\frac{1}{m(I)} \int_I |f - \frac{1}{m(I)} \int_I f dm|^p dm = - \frac{1}{m(I)} \int_0^\infty t^p d\lambda_I(t)$$

$$= - \frac{1}{m(I)} [t^p \lambda_I(t)|_0^\infty - \int_0^\infty p t^{p-1} \lambda_I(t)dt]$$

$$\leq \frac{1}{m(I)} \int_0^\infty p t^{p-1} \lambda_I(t)dt = \frac{1}{m(I)} [\int_0^N p t^{p-1} \lambda_I(t)dt + \int_N^\infty p t^{p-1} \lambda_I(t)dt]$$

$$\leq \frac{1}{m(I)} \left[m(I) \int_0^N pt^{p-1} dt + m(I) \int_N^\infty pt^{p-1} e^{-bt} dt\right]$$

$$= \int_0^N pt^{p-1} dt + \int_N^\infty pt^{p-1} e^{-bt} dt.$$

Since the final expression is a finite constant independent of I, the proof is complete.

This Corollary shows that there is no need to be alarmed at the squares (present only to facilitate some of the computations) which appear in several of the characterizations of BMO.

Lemma 7.2.4. Suppose that $f \in L^2(\Gamma)$, and for each short interval $I \subset \Gamma$ there is a number α_I such that

$$\sup_I \frac{1}{m(I)} \int_I |f(e^{i\sigma}) - \alpha_I|^2 dm(\sigma) < \infty.$$

Then f has bounded mean oscillation.

Proof. For each interval $I \subset \Gamma$, let

$$f_I = \frac{1}{m(I)} \int_I f dm,$$

and denote the hypothetical finite supremum by A. Then for each short interval I,

$$\left[\frac{1}{m(I)} \int_I |f - f_I|^2 dm\right]^{\frac{1}{2}} = \frac{1}{\sqrt{(m(I))}} \|f - f_I\|_{L^2(I)}$$

$$\leq \frac{1}{\sqrt{(m(I))}} \left[\|f - \alpha_I\|_{L^2(I)} + \|\alpha_I - f_I\|_{L^2(I)}\right]$$

$$\leq \sqrt{A} + |\alpha_I - f_I| = \sqrt{A} + \frac{1}{m(I)} \left[\int_I (f - \alpha_I) dm\right]$$

$$\leq \sqrt{A} + \frac{1}{m(I)} \int_I |f - \alpha_I| dm \leq \sqrt{A} + \frac{1}{m(I)} \|f - \alpha_I\|_{L^2(I)} \|1\|_{L^2(I)}$$

$$\leq \sqrt{A} + \frac{1}{m(I)} \sqrt{A} \sqrt{(m(I))} \sqrt{(m(I))} = 2\sqrt{A},$$

and $f \in BMO$ by Corollary 7.2.3.

The following Lemma contains the most difficult step in our sequence of implications.

Lemma 7.2.5. *Let $\phi \in$ BMO; there is a constant C_ϕ such that if $0 < r < 1$, then there is a subinterval I of Γ, centered at 1, such that*

$$\int_\Gamma \left|\phi - \frac{1}{m(I)} \int_I \phi dm\right|^2 \mathcal{P}(r, \psi) dm(\psi) \le C_\phi .$$

Proof. Let $I_n = \{e^{i\theta} : -\frac{\pi}{2^n} \le \theta \le \frac{\pi}{2^n}\}$ for $n = 0, 1, 2, \ldots$, so that $m(I_n) = \frac{1}{2^n}$ and $I_{n+1} \subset I_n$. Let

$$\phi_k = \frac{1}{m(I_k)} \int_{I_k} \phi dm ,$$

and fix a large value of n. Let

$$A = \sup_I \frac{1}{m(I)} \int_I \left|\phi - \frac{1}{m(I)} \int_I \phi dm\right|^2 dm ;$$

$A < \infty$ by 7.2.3. Note that

$$|\phi_{k+1} - \phi_k| = \left|\frac{1}{m(I_{k+1})} \int_{I_{k+1}} (\phi - \phi_k) dm\right| \le \frac{2}{m(I_k)} \int_{I_k} |\phi - \phi_k| dm \le 2\|\phi\|_{BMO} ,$$

and for $0 \le k < n$,

$$\left[\int_{I_k \setminus I_{k+1}} |\phi - \phi_n|^2 dm\right]^{\frac{1}{2}} \le \left[\int_{I_k \setminus I_{k+1}} |\phi - \phi_k|^2 dm\right]^{\frac{1}{2}} + \sum_{j=k}^{n-1} \left[\int_{I_k \setminus I_{k+1}} |\phi_j - \phi_{j+1}|^2 dm\right]^{\frac{1}{2}}$$

$$\le \sqrt{(A/2^k)} + 2\|\phi\|_{BMO} \sum_{j=k}^{n-1} \sqrt{(m(I_k \setminus I_{k+1}))} + \sqrt{(D/2^k)}\,(n - k)$$

for some constant D which depends on ϕ. We are trying to estimate

$$\int_\Gamma |\phi - \phi_n|^2 \mathcal{P}(r, \psi) dm(\psi) = \sum_{k=0}^{n-1} \int_{I_k \setminus I_{k+1}} |\phi - \phi_n|^2 \mathcal{P}(r, \psi) dm(\psi) + \int_{I_n} |\phi - \phi_n|^2 \mathcal{P}(r, \psi) dm(\psi)$$

For fixed r with $0 \le r < 1$, let $\delta = 1 - r$ and choose n so that $\frac{1}{2^n} \le \delta < \frac{1}{2^{n-1}}$. By obvious estimates,

$$\mathcal{P}(r, \psi) \le \frac{1 + r}{1 - r} \le \frac{2}{\delta} \le \frac{2}{m(I_n)} ,$$

so that the second term is no larger than $2A$; and since there are constants a and b such that

$$\mathcal{P}(r, \psi) \le \frac{a\delta}{\delta^2 + b\left(\frac{\psi}{\pi}\right)^2}$$

for all $\psi \in [-\pi, \pi]$, the first term is no larger than

$$\sum_{k=0}^{n-1} \int_{I_k \setminus I_{k+1}} |\phi - \phi_n|^2 \frac{a\delta}{\delta^2 + b(\frac{1}{2^{k+1}})^2} dm(\psi) \leq \sum_{k=0}^{n-1} \frac{\frac{a}{2^{n-1}}}{\frac{1}{2^{2n}} + \frac{b}{2^{2k+2}}} \frac{D(n-k)^2}{2^k}$$

$$= 2aD \sum_{k=0}^{n-1} \frac{(n-k)^2 2^{n-k}}{1 + b2^{2(n-k-1)}}$$

$$\leq \frac{8aD}{b} \sum_{j=1}^{\infty} \frac{j^2}{2^j} = K < \infty .$$

Thus we may take $C_\phi = K + 2A$.

Proposition 7.2.6. <u>Let</u> $\phi \in L^2_R(\Gamma)$. <u>Then</u> $\phi \in BMO$ <u>if and only if</u>

$$\sup_{r, \theta} \int_\Gamma [\phi(e^{i\sigma}) - \int_\Gamma \phi(e^{i\psi}) \mathcal{P}(r, \theta-\psi) dm(\psi)]^2 \mathcal{P}(r, \theta-\sigma) dm(\sigma) < \infty .$$

Proof. Suppose that ϕ satisfies this condition. We will convert the Poisson averages to interval averages by associating a suitable interval length $2\sigma_r$ with each $r \in [0, 1)$: for each such r we choose $\sigma_r \in (0, \pi)$ such that

$$\mathcal{P}(r, \sigma_r) = \tfrac{1}{2}[\frac{1+r}{1-r} + \frac{1-r}{1+r}] = \frac{1 + r^2}{1 - r^2},$$

so that $\pm\sigma_r$ are the points at which $\mathcal{P}(r, \psi)$ assumes the average of its maximum and minimum values. Let

$$V_r(\psi) = \frac{1}{2\sigma_r} \chi_{[-\sigma_r, \sigma_r]}(\psi) \quad \text{for} \quad -\pi \leq \psi \leq \pi.$$

I claim that there is a constant c independent of r such that $V_r \leq c\mathcal{P}_r$ on $[-\pi, \pi]$. For this purpose it is enough to show that

$$2\sigma_r \frac{1 + r^2}{1 - r^2}$$

is bounded below for $0 \leq r < 1$, for then it will follow that

$$V_r \leq \frac{1}{2\sigma_r} \leq c \frac{1 + r^2}{1 - r^2} \leq c\mathcal{P}_r.$$

Now if $r = 0$ we may take $\sigma_r = 1$, while for $r > 0$ the number σ_r is determined by the equation

$$\frac{1 - r^2}{1 - 2r \cos \sigma_r + r^2} = \frac{1 + r^2}{1 - r^2} ,$$

which implies that

$$\sin^2 \sigma_r = \left(\frac{1 - r^2}{1 + r^2}\right)^2 .$$

Therefore for $0 \leq r < 1$,

$$2\sigma_r \frac{1 + r^2}{1 - r^2} \geq 2 \sin \sigma_r \frac{1 + r^2}{1 - r^2} = 2,$$

proving our assertion.

Now given any interval $I \subset \Gamma$, let $e^{i\theta}$ denote the midpoint of I and choose r so that $m(I) = 2\sigma_r$. Then

$$V_r(\theta - \sigma) = \frac{1}{m(I)} \chi_I(\sigma) \quad \text{for} \quad \sigma \in [-\pi, \pi].$$

Let $\alpha_I = \int_\Gamma \phi(e^{i\psi}) \mathcal{P}(r, \theta - \psi) dm(\psi)$. Then

$$\frac{1}{m(I)} \int_I [\phi - \alpha_I]^2 dm = \int_\Gamma [\phi(e^{i\sigma}) - \alpha_I]^2 V_r(\theta - \sigma) dm(\sigma)$$

$$\leq c \int_\Gamma [\phi(e^{i\sigma}) - \alpha_I]^2 \mathcal{P}(r, \theta - \sigma) dm(\sigma),$$

which by hypothesis is bounded by a constant independent of I. By Lemma 7.2.4, then, $\phi \in \mathrm{BMO}$.

Conversely, suppose now that $\phi \in \mathrm{BMO}$. Given $z = re^{i\theta} \in D$, choose an interval of the type whose existence is guaranteed by Lemma 7.2.5 and let I be the translate of this interval which is centered at $e^{i\theta}$. Let us abbreviate

$$\phi_I = \frac{1}{m(I)} \int_I \phi \, dm \quad \text{and}$$

$$\mathcal{P}\phi(z) = \int_\Gamma \phi(e^{i\psi}) \mathcal{P}(r, \theta - \psi) dm(\psi) .$$

Then

$$\left[\int_\Gamma [\phi(e^{i\sigma}) - \mathcal{P}\phi(z)]^2 \mathcal{P}(r, \theta - \sigma) dm(\sigma)\right]^{\frac{1}{2}}$$

$$\leq \left[\int_\Gamma [\phi - \phi_I]^2 \mathcal{P}(r, \theta - \sigma) dm(\sigma)\right]^{\frac{1}{2}} + \left[\int_\Gamma [\phi_I - \mathcal{P}\phi(z)]^2 \mathcal{P}(r, \theta - \sigma) dm(\sigma)\right]^{\frac{1}{2}}.$$

The first term is less than or equal to $\sqrt{C_\phi}$, while in the second

$$|\phi_I - \mathcal{P}\phi(z)| \leq \int_\Gamma |\phi_I - \phi(e^{i\psi})| \mathcal{P}(r, \theta - \psi) dm(\psi)$$

$$\leq \left[\int_\Gamma |\phi_I - \phi(e^{i\psi})|^2 \mathcal{P}(r, \theta - \psi) dm(\psi)\right]^{\frac{1}{2}} \leq \sqrt{C_\phi} ,$$

so that the entire expression is bounded by $2\sqrt{C_\phi}$ independently of r and θ.

The power of this result is apparent in the ease with which it yields the following Corollary, which is sometimes called the Theorem on the Conjugate Function.

Corollary 7.2.7. *Let $\phi \in L^2_R(\Gamma)$ and let $\tilde{\phi}$ be its Fourier conjugate. If ϕ has bounded mean oscillation, then so does $\tilde{\phi}$.*

Proof. Let $u = \mathcal{P}\phi$, so that $\tilde{u} = \mathcal{P}\tilde{\phi}$, and let $F = u + i\tilde{u}$. For fixed $z = re^{i\theta} \in D$, we employ a linear fractional transformation

$$T\zeta = \frac{\zeta - z}{1 - \bar{z}\zeta} \quad \text{with inverse} \quad T^{-1}\zeta = \frac{\zeta + z}{1 + \bar{z}\zeta}$$

and define Q_z on Γ by

$$Q_z(e^{i\psi}) = F\left(\frac{e^{i\psi} + z}{1 + \bar{z}e^{i\psi}}\right) - F(z) .$$

Since

$$\left| \frac{d}{d\psi}\left(\frac{e^{i\psi} - z}{1 - \bar{z}e^{i\psi}}\right) \right| = \mathcal{P}(r,\ \theta - \psi),$$

we have

$$\frac{1}{2\pi} \int_\Gamma [\mathrm{Re}Q_z(e^{i\psi})]^2 d\psi = \frac{1}{2\pi} \int_\Gamma [\phi(e^{i\psi}) - u(z)]^2 \mathcal{P}(r,\ \theta - \psi) d\psi .$$

By Riesz' Inequalities (2.5 and 5.2.3), then, $Q_z(\zeta)$ is in H^2 and

$$\|Q_z\|^2_{H^2} \le c \frac{1}{2\pi} \int_\Gamma [\phi(e^{i\psi}) - u(z)]^2 \mathcal{P}(r,\ \theta - \psi) d\psi,$$

which by hypothesis and Proposition 7.2.6 is bounded by a constant independent of z. Therefore

$$\sup_{z \in D} \frac{1}{2\pi} \int_\Gamma [\tilde{\phi}(e^{i\psi}) - \tilde{u}(z)]^2 \mathcal{P}(r,\ \theta - \psi) d\psi \le \sup_{z \in D} \frac{1}{2\pi} \int_\Gamma |F(e^{i\psi}) - F(z)|^2 \mathcal{P}(r, \theta - \psi) d\psi$$

$$= \sup_{z \in D} \|Q_z\|^2_{H^2} < \infty,$$

so that $\tilde{\phi} \in \mathrm{BMO}$.

We may rewrite the significant expression which appears in Proposition 7.2.6 in a slightly more compact notation. Continue to denote the Poisson integral of a function f on Γ by $\mathcal{P}f$, so that for $z = re^{i\theta} \in D$,

$$\mathcal{P}f(z) = \frac{1}{2\pi} \int_0^{2\pi} f(e^{i\psi}) \mathcal{P}(r, \theta - \psi) d\psi.$$

The expression appearing in 7.2.6 and (3) may be written, then, as

$$\sup_{z \in D} \frac{1}{2\pi} \int_\Gamma [\phi(e^{i\sigma}) - \mathcal{P}\phi(z)]^2 \mathcal{P}(r, \theta - \sigma) d\sigma$$

$$= \sup_{z \in D} \{[\mathcal{P}(\phi^2)(z)] - [\mathcal{P}\phi(z)]^2\}$$

$$= \sup_{z \in D} [\mathcal{P}\phi^2(z) - \mathcal{P}\phi(z)^2].$$

Remark 7.2.8. Since a complex-valued function f on Γ is in BMO if and only if its real and imaginary parts are, Proposition 7.2.6 implies that $f \in L^2(\Gamma)$ has bounded mean oscillation if and only if

$$\sup_{z \in D} \frac{1}{2\pi} \int_\Gamma |f(e^{i\sigma}) - F(z)|^2 \mathcal{P}(r, \theta - \sigma) d\sigma < \infty,$$

where $z = re^{i\theta}$ and $F = \mathcal{P}f$. This is equivalent to the condition that

$$\sup_{z \in D} (\mathcal{P}|f|^2(z) - |\mathcal{P}f(z)|^2) < \infty.$$

On the basis of Proposition 7.2.6, we can now proceed to give two probabilistic characterizations of bounded mean oscillation. As in earlier chapters, denote by $\tilde{\gamma}_{z,t}(\omega)$ the Brownian motion starting at $z \in D$ which is absorbed at its first point of impact on Γ, at time $\tau_z(\omega) = \inf\{t \geq 0 : |\gamma_{z,t}(\omega)| \geq 1\}$.

Proposition 7.2.9. <u>A function</u> $\phi \in L^2_{\mathbf{R}}(\Gamma)$ <u>is in BMO if and only if</u>

$$\sup_{z \in D} \int_\Omega [\phi(\gamma_{z,\tau}) - u(z)]^2 dP < \infty,$$

<u>where</u> $u = \mathcal{P}\phi$ <u>is the harmonic function on</u> D <u>with boundary function</u> ϕ.

Proof. Immediate from Proposition 7.2.6 and Corollary 3.6.2.

Remark 7.2.10. Let $\phi \in L^2_{\mathbf{R}}(\Gamma)$, $u = \mathcal{P}\phi$, and $F = u + i\tilde{u}$. Let us suppress ϕ and denote the boundary functions of u and F by u and F, respectively. Since $u(\tilde{\gamma}_t)$ is measurable with respect to $\tilde{\mathcal{B}}_t$ and $\mathcal{B}(\tilde{\gamma}_t)$, an easy calculation with 7.1 (1) shows that

$$E([u(\tilde{\gamma}_{s+t}) - u(\tilde{\gamma}_t)]^2 | \tilde{\mathcal{B}}_t) = E([u(\tilde{\gamma}_{s+t}) - u(\tilde{\gamma}_t)]^2 | \tilde{\gamma}_t) \text{ a. e. ;}$$

of course this implies that also

$$E(|F(\tilde{\gamma}_{s+t}) - F(\tilde{\gamma}_t)|^2 | \tilde{\mathcal{B}}_t) = E(|F(\tilde{\gamma}_{s+t}) - F(\tilde{\gamma}_t)|^2 | \tilde{\gamma}_t) \text{ a. e.}$$

By the martingale property (see 7.1), these expressions may be re-written as

$$E(u(\tilde{\gamma}_{s+t})^2 - u(\tilde{\gamma}_t)^2 | \tilde{\gamma}_t) \text{ and}$$

$$E(|F(\tilde{\gamma}_{s+t})|^2 - |F(\tilde{\gamma}_t)|^2 | \tilde{\gamma}_t),$$

respectively.

For each $z \in D$ and each measurable set $G \subset D$, let

$$\mu_{z,t}(G) = P\{\omega : \tilde{\gamma}_{z,t}(\omega) \in G\}.$$

Since

$$\mu_{z,t}(G) \leq P\{\omega : \gamma_{z,t}(\omega) \in G\} = \frac{1}{2\pi t} \iint_G e^{-r^2/2t} r dr d\theta,$$

$\mu_{z,t}$ is absolutely continuous with respect to two-dimensional Lebesgue measure dA, and hence there is a density function

$$\delta(z, t, w) \leq \frac{1}{2\pi t} e^{-|z-w|^2/2t}$$

such that

$$\mu_{z,t}(G) = \iint_G \delta(z, t, w) dA(w).$$

We will denote $\mu_{0,t}$ by μ_t.

Proposition 7.2.11. <u>Let</u> $\phi \in L^2_R(\Gamma)$, $u = \mathcal{P}\phi$ <u>on</u> D, <u>and</u> $u = \phi$ <u>on</u> Γ. <u>Then</u> $\phi \in$ BMO <u>if and only if</u>

$$\sup_{t \geq 0} \| E(u(\gamma_\tau)^2 - u(\tilde{\gamma}_t)^2 | \tilde{\gamma}_t) \|_{L^\infty(\Omega)} < \infty.$$

Proof. We use Remark 7.2.10 and note that

$$\sup_{t \geq 0} \| E([u(\gamma_\tau) - u(\tilde{\gamma}_t)]^2 | \tilde{\gamma}_t) \|_{L^\infty(\Omega)} \leq K < \infty$$

if and only if for each measurable $G \subset D$

$$\sup_{t\geq 0} \frac{1}{\mu_t(G)} \iint_{\tilde{\gamma}_t^{-1}G} E([u(\gamma_\tau) - u(\tilde{\gamma}_t)]^2 \,|\, \tilde{\gamma}_t) dP \leq K .$$

By Corollary 3.5.4 and the preceding remarks, the latter expression equals

$$\sup_{t\geq 0} \frac{1}{\mu_t(G)} \iint_G \left(\int_\Omega [u(\tilde{\gamma}_{z,\tau}) - u(z)]^2 dP \right) d\mu_t(z),$$

and of course this is no greater than K if and only if

$$\sup_{z\in D} \int_\Omega [u(\tilde{\gamma}_{z,\tau}) - u(z)]^2 dP \leq K ,$$

since the Brownian paths $\tilde{\gamma}_t(\omega)$ enter each open subset of D with positive probability. Proposition 7.2.9 then yields the desired conclusion.

We can give further characterizations of bounded mean oscillation, involving the <u>Green's function</u>

$$g(z, w) = \log\left|\frac{1 - \bar{w}z}{z - w}\right| \qquad (z, w \in D)$$

of D, by appealing to a simple computation with power series and a remarkable theorem of Hunt's.

Lemma 7.2.12. <u>If</u> $Q \in H^2$ <u>and</u> $Q(0) = 0$, <u>then</u>

$$\frac{1}{2\pi} \int_0^{2\pi} |Q(e^{i\psi})|^2 d\psi = \frac{2}{\pi} \iint_D |Q'(w)|^2 \log(\frac{1}{|w|}) dA(w).$$

Proof. Let $Q(w) = \sum_{n=1}^{\infty} a_n w^n$, so that

$$\frac{1}{2\pi} \int_0^{2\pi} |Q(e^{i\psi})|^2 d\psi = \sum_{n=1}^{\infty} |a_n|^2.$$

Then, writing $w = \rho e^{i\sigma}$,

$$\iint_D |Q'(w)|^2 \log(\frac{1}{|w|}) dA(w) = -\iint_D \sum_{i,j=1}^{\infty} a_i \bar{a}_j ij \rho^{i+j-2} e^{(i-j)\sigma} (\log \rho) \rho \, d\rho \, d\sigma$$

$$= -2\pi \sum_{n=1}^{\infty} n^2 |a_n|^2 \int_0^1 \rho^{2n-1} \log \rho \, d\rho = -2\pi \sum_{n=1}^{\infty} n^2 |a_n|^2 (-\frac{1}{4n^2})$$

$$= \frac{\pi}{2} \sum_{n=1}^{\infty} |a_n|^2 .$$

Proposition 7.2.13. <u>The boundary function of an</u> H^2 <u>function</u> F <u>has bounded mean oscillation if and only if</u>

$$\sup_{z \in D} \iint_D |F'(w)|^2 g(z, w)\, dA(w) < \infty,$$

<u>where</u> $g(z, w) = \log\left|\frac{1 - \bar{w}z}{z - w}\right|$ <u>is the Green's function of</u> D.

Proof. As in the proof of 7.2.7, for fixed $z \in D$ let

$$Q_z(w) = F\left(\frac{w + z}{1 + \bar{z}w}\right) - F(z) \qquad (w \in \bar{D}).$$

Then Q_z is analytic on D, $Q_z(0) = 0$, and

$$\frac{1}{2\pi}\int_0^{2\pi} |Q_z(e^{i\psi})|^2 d\psi = \frac{1}{2\pi}\int_0^{2\pi} |F(e^{i\psi}) - F(z)|^2 \wp(r, \theta-\psi)\, d\psi,$$

so that $Q_z \in H^2$. Of course

$$Q_z'(w) = F'\left(\frac{w + z}{1 + \bar{z}w}\right) \frac{d}{dw}\left(\frac{w + z}{1 + \bar{z}w}\right);$$

but if we let

$$v = \frac{w + z}{1 + \bar{z}w},$$

then the Jacobian of the mapping $w \to v$ is

$$\left|\frac{d}{dw}\left(\frac{w + z}{1 + \bar{z}w}\right)\right|^2,$$

and hence

$$\frac{2}{\pi}\iint_D |Q_z'(w)|^2 \log\left(\frac{1}{|w|}\right) dA(w) = \frac{2}{\pi}\iint_D \left|F'\left(\frac{w + z}{1 + \bar{z}w}\right)\right|^2 \left|\frac{d}{dw}\left(\frac{w + z}{1 + \bar{z}w}\right)\right|^2 \log\left(\frac{1}{|w|}\right) dA(w)$$

$$= \frac{2}{\pi}\iint_D |F'(v)|^2 \log\left|\frac{1 - \bar{z}v}{z - v}\right| da(v) = \frac{2}{\pi}\iint_D |F'(v)|^2 g(z, v)\, dA(v).$$

By Lemma 7.2.12, then,

$$\frac{2}{\pi}\iint_D |F'(v)|^2 g(z, v)\, dA(v) = \frac{1}{2\pi}\int_0^{2\pi} |Q_z(e^{i\psi})|^2 d\psi$$

$$= \frac{1}{2\pi}\int_0^{2\pi} |F(e^{i\psi}) - F(z)|^2 \wp(r, \theta - \psi)\, d\psi,$$

and according to Remark 7.2.8 this is a bounded function of $z \in D$ if and only if the boundary function of F is in BMO.

Theorem 7.2.14 (Hunt [26]; see also [27, §7.4]). *If* $\delta(z, t, w)$ *is defined for* $z, w \in D$ *and* $t \geq 0$ *by*

$$P\{\omega : \tilde{\gamma}_{z,t}(\omega) \in G\} = \iint_G \delta(z, t, w)\,dA(w)$$

for each measurable $G \subset D$, *then*

$$\int_0^\infty \delta(z, t, w)\,dt = g(z, w) = \log\left|\frac{1 - \bar{w}z}{z - w}\right|$$

for almost all $z, w \in D$.

We omit the somewhat difficult proof of this result, noting only that the central argument uses the strong Markov property, as in the proof of Theorem 3.6.1, to establish that $\int_0^\infty \delta(z, t, w)\,dt$ is equal almost everywhere to a function harmonic in both z and w.

Proposition 7.2.15. *The boundary function of an* H^2 *function* F *has bounded mean oscillation if and only if*

$$\sup_{t\geq 0} \left\| E\left(\int_t^\tau |F'(\tilde{\gamma}_s)|^2 ds \,\middle|\, \tilde{\gamma}_t\right) \right\|_{L^\infty(\Omega)} < \infty.$$

Proof. Let us note that by Theorem 7.2.14 and the remarks preceding Proposition 7.2.11, for fixed $z \in D$

$$\iint_D |F'(w)|^2 g(z, w)\,dA(w) = \iint_D |F'(w)|^2 \int_0^\infty \delta(z, t, w)\,dt\,dA(w)$$

$$= \int_0^\infty \left[\iint_D |F'(w)|^2 \delta(z, t, w)\,dA(w)\right]dt = \int_0^\infty \left[\iint_D |F'(w)|^2 d\mu_{z,t}(w)\right]dt$$

$$= \int_0^\infty \left[\int_{\{t<\tau\}} |F'(\tilde{\gamma}_{z,t})|^2 dP\right]dt = \int_\Omega \int_0^{\tau(\omega)} |F'(\tilde{\gamma}_{z,t}(\omega))|^2 dt\,dP(\omega).$$

Now if $G \subset D$ is measurable, Corollary 3.5.4 allows us to write

$$\int_{\tilde{\gamma}_t^{-1}G} E\left(|F(\gamma_\tau) - F(\tilde{\gamma}_t)|^2 \,\middle|\, \tilde{\gamma}_t\right)dP = \int_{\tilde{\gamma}_t^{-1}G} E\left(|F(\gamma_{\tilde{\gamma}_t(\omega),\tau}) - F(\tilde{\gamma}_t(\omega))|^2\right)dP(\omega)$$

$$= \iint_G \left[\int_\Omega |F(\gamma_{z,\tau}) - F(z)|^2 dP\right]d\mu_t(z) = \iint_G \left[\int_\Omega |F(e^{i\sigma}) - F(z)|^2 \mathcal{P}(r, \theta-\sigma)\,dm(\sigma)\right]d\mu_t(z)$$

$$= \frac{2}{\pi} \iint_G \left[\iint_D |F'(w)|^2 g(z, w)\,dA(w)\right]d\mu_t(z)$$

$$= \frac{2}{\pi} \iint_G \left[\int_\Omega \int_0^{\tau(\omega)} |F'(\tilde{\gamma}_{z,s}(\omega))|^2 ds\,dP(\omega)\right]d\mu_t(z),$$

where we have also used the proof of Proposition 7. 2. 13 and our preceding remarks. Now we may unravel this computation as follows:

$$\frac{2}{\pi}\iint_G [\int_\Omega \int_0^{\tau(\omega)} |F'(\tilde{\gamma}_{z,s}(\omega))|^2 ds dP(\omega)] d\mu_t(z)$$

$$= \frac{2}{\pi}\iint_G E(\int_0^\tau |F'(\tilde{\gamma}_{z,s})|^2 ds) d\mu_t(z)$$

$$= \frac{2}{\pi}\iint_G [\int_0^\infty E(\chi_{\{s<\tau\}} |F'(\tilde{\gamma}_{z,s})|^2) ds] d\mu_t(z)$$

$$= \frac{2}{\pi}\iint_G [\int_0^\infty E(\chi_D(\tilde{\gamma}_{z,s}) |F'(\tilde{\gamma}_{z,s})|^2) ds] d\mu_t(z)$$

$$= \frac{2}{\pi}\int_{\tilde{\gamma}_t^{-1}G} [\int_0^\infty E(\chi_D(\tilde{\gamma}_{s+t}) |F'(\tilde{\gamma}_{s+t})|^2 \,|\, \tilde{\mathcal{B}}_t) ds] dP$$

$$= \frac{2}{\pi}\int_0^\infty [\int_{\tilde{\gamma}_t^{-1}G} \chi_D(\tilde{\gamma}_{s+t}) |F'(\tilde{\gamma}_{s+t})|^2 dP] ds$$

$$= \frac{2}{\pi}\int_{\tilde{\gamma}_t^{-1}G} [\int_t^{\tau(\omega)} |F'(\tilde{\gamma}_s(\omega))|^2 ds] dP(\omega),$$

and hence

$$E(|F(\gamma_\tau) - F(\tilde{\gamma}_t)|^2 \,|\, \tilde{\gamma}_t) = \frac{2}{\pi} E(\int_t^\tau |F'(\tilde{\gamma}_s)|^2 ds \,|\, \tilde{\gamma}_t) \quad \text{a. e.}$$

Then the conclusion follows from Proposition 7. 2. 11.

Remarks 7. 2. 16. Let $u = \operatorname{Re} F$, so that $|F'|^2 = |\nabla u|^2$. Note that 7. 1 (1) can be extended to show that

$$E\left(\int_t^\tau |\nabla u(\tilde{\gamma}_r)|^2 dr \,\middle|\, \tilde{\gamma}_t\right) = E\left(\int_t^\tau |\nabla u(\tilde{\gamma}_r)|^2 dr \,\middle|\, \tilde{\mathcal{B}}_t\right).$$

By the proof of Corollary 7. 2. 7, Proposition 7. 2. 15, and Remark 7. 2. 10, if $0 \le s \le t$ then a. e. on $\{t < \tau\}$ we have

$$E\left(\int_s^t |\nabla u(\tilde{\gamma}_r)|^2 dr \,\middle|\, \tilde{\gamma}_s\right) = E\left(\int_s^\tau |\nabla u(\tilde{\gamma}_r)|^2 dr \,\middle|\, \tilde{\gamma}_s\right) - E\left(\int_t^\tau |\nabla u(\tilde{\gamma}_r)|^2 dr \,\middle|\, \tilde{\gamma}_s\right)$$

$$= \frac{\pi}{2}[E(u(\gamma_\tau)^2 - u(\tilde{\gamma}_s)^2 \,|\, \tilde{\gamma}_s) + E(\tilde{u}(\gamma_\tau)^2 - \tilde{u}(\tilde{\gamma}_s)^2 \,|\, \tilde{\gamma}_s)$$

$$- E(E(u(\gamma_\tau)^2 - u(\tilde{\gamma}_t)^2 \,|\, \tilde{\gamma}_t) \,|\, \tilde{\gamma}_s) - E(E(\tilde{u}(\gamma_\tau)^2 - \tilde{u}(\tilde{\gamma}_t)^2 \,|\, \tilde{\gamma}_t) \,|\, \tilde{\gamma}_s)]$$

$$= \frac{\pi}{2}[E(u(\tilde{\gamma}_t)^2 - u(\tilde{\gamma}_s)^2 \,|\, \tilde{\gamma}_s) + E(\tilde{u}(\tilde{\gamma}_t)^2 - \tilde{u}(\tilde{\gamma}_s)^2 \,|\, \tilde{\gamma}_s)]$$

$$\le c\, E(u(\tilde{\gamma}_t)^2 - u(\tilde{\gamma}_s)^2 \,|\, \tilde{\gamma}_s).$$

Corollary 7.2.17. Let $\phi \in L^2_{\mathbf{R}}(\Gamma)$ and $u = \mathcal{P}\phi$. Then $\phi \in$ BMO if and only if

$$\sup_{t\geq 0} \left\| E\left(\int_t^\tau |\nabla u(\tilde{\gamma}_s)|^2 ds \,\middle|\, \tilde{\gamma}_t \right) \right\|_{L^\infty(\Omega)} < \infty .$$

The justification of the definition of BMO martingales (Definition 7.2.1) can now be given in terms of Proposition 7.2.11. For if we are given $\phi \in L^2_{\mathbf{R}}(\Gamma)$ and let $u = \mathcal{P}\phi$ and

$$x_t = u(\tilde{\gamma}_t) \text{ for } t \geq 0,$$

then $\{x_t\}$ is an H^2 martingale with respect to $\{\tilde{\mathcal{B}}_t\}$ and converges a.e. to $x_\infty = \phi(\gamma_\tau)$. Of course x_∞ is measurable with respect to $\mathcal{B}(\gamma_\tau)$ and $x_t = E(x_\infty | \tilde{\mathcal{B}}_t)$ a.e. By Proposition 7.2.11,

$$\sup_{t\geq 0} \| E(|x_\infty - x_t|^2 | \tilde{\mathcal{B}}_t) \|_{L^\infty(\Omega)} < \infty$$

if and only if ϕ has bounded mean oscillation.

Conversely, if $\{x_t\}$ is an H^2 martingale with respect to $\{\tilde{\mathcal{B}}_t\}$ which satisfies this condition, and for which x_∞ is measurable with respect to $\mathcal{B}(\gamma_\tau)$, then we may argue as in the preceding section that there is $\phi \in L^2_{\mathbf{R}}(\Gamma)$ such that $x_\infty = \phi(\gamma_\tau)$ a.e., and it also follows then that $x_t = E(x_\infty | \tilde{\mathcal{B}}_t) = u(\tilde{\gamma}_t)$ a.e. Then u must satisfy the condition of Proposition 7.2.11, and therefore ϕ has bounded mean oscillation.

The family of all real-valued BMO functions on Γ is identified in this way with the family of all real-valued martingales $\{x_t\}$ on $(\Omega, \mathcal{B}, P)$ which satisfy (1) and have, in addition, x_∞ measurable with respect to $\mathcal{B}(\gamma_\tau)$. The condition stated in Corollary 7.2.17 could also serve as the basis for a definition of BMO martingales in the abstract setting (using the notion of 'the increasing process associated with a martingale' ([44, p. 164]), and in fact some authors (e.g. [23]) do take this as their starting point.

We remind the reader that a list of the different characterizations of bounded mean oscillation mentioned in these pages can be found in 7.3.3.

7.3 The duality theorem

The ideas developed in the preceding two sections can be used to give a proof of the Fefferman-Stein Theorem on the duality of H^1 and BMO (2.11). The martingale and analytic versions of this theorem have been greatly extended in generality and scope by several authors; the probabilistic proof of the basic result given here is intended to illuminate once again the relationship between these two points of view. The following Lemma is one of the Burgess Davis inequalities (2.7; see [7] and [23] for the continuous-parameter case), while the names Fefferman, Gundy, Herz, Stein, and Garsia should be associated with Theorem 7.3.2 and its proof.

Lemma 7.3.1. *For each* $F \in H^2$ *and* $\omega \in \Omega$ *define*

$$SF(\omega) = [\int_0^{\tau(\omega)} |F'(\tilde{\gamma}_t)|^2 dt]^{\frac{1}{2}} .$$

There is a constant c *such that whenever* $F \in H^2$ *and* $F(0) = 0$,

$$\|SF\|_{L^1(\Omega)} \leq c \|F^*\|_{L^1(\Omega)} .$$

Proof. For each $t \geq 0$, let

$$S_t(\omega) = [\int_0^{t \wedge \tau(\omega)} |F'(\tilde{\gamma}_s)|^2 ds]^{\frac{1}{2}} \quad \text{and}$$

$$F_t^*(\omega) = \sup_{0 \leq s \leq t} |F(\tilde{\gamma}_s(\omega))| .$$

Fix $\varepsilon > 0$ and write

$$\int_\Omega S_t dP = \int_\Omega \sqrt{(F_t^* + \varepsilon)} \frac{S_t}{\sqrt{(F_t^* + \varepsilon)}} dP$$
$$\leq \left(\int_\Omega (F_t^* + \varepsilon) dP \right)^{\frac{1}{2}} \left(\int_\Omega \frac{S_t^2}{F_t^* + \varepsilon} dP \right)^{\frac{1}{2}} .$$

Here the first factor is bounded by $\sqrt{(\|F^* + \varepsilon\|_1)}$ while the square of the second equals $E\left(\frac{S_t^2}{F_t^* + \varepsilon}\right)$.

Let us restrict our attention now to the probability subspace $\Omega_t = \{t < \tau\}$ of $(\Omega, \mathcal{B}, P)$. Take a partition $\pi : 0 = t_0 < t_1 < \ldots < t_n = t$

of $[0, t)$ and abbreviate $S_k = S_{t_k}$, $F^*_k = F^*_{t_k}$, and $\mathcal{F}_k = \tilde{\mathcal{B}}_{t_k}$. Then we may write

$$\frac{S_t^2}{F_t^* + \varepsilon} = \sum_{k=0}^{n-1}\left(\frac{S_{k+1}^2}{F_{k+1}^* + \varepsilon} - \frac{S_k^2}{F_k^* + \varepsilon}\right) \leq \sum_{k=0}^{n-1} \frac{S_{k+1}^2 - S_k^2}{F_k^* + \varepsilon},$$

and taking conditional expectations in turn with respect to $\mathcal{F}_{n-1}, \mathcal{F}_{n-2}, \ldots, \mathcal{F}_0 = \{\emptyset, \Omega\}$ gives, via Remark 7.2.16 with $u = \operatorname{Re} F$,

$$E_{\Omega_t}\left(\frac{S_t^2}{F_t^* + \varepsilon}\,\Big|\,\mathcal{F}_{n-1}\right) \leq \frac{E_{\Omega_t}(S_n^2 - S_{n-1}^2 \,|\, \mathcal{F}_{n-1})}{F_{n-1}^* + \varepsilon} + \sum_{k=0}^{n-2} \frac{S_{k+1}^2 - S_k^2}{F_k^* + \varepsilon}$$

$$= \frac{1}{F_{n-1}^* + \varepsilon} E_{\Omega_t}\left(\int_{t_{n-1}}^{t_n} |\nabla u(\tilde{\gamma}_s)|^2 ds \,\Big|\, \mathcal{F}_{n-1}\right) + \sum_{k=0}^{n-2} \frac{S_{k+1}^2 - S_k^2}{F_k^* + \varepsilon}$$

$$\leq \frac{c}{F_{n-1}^* + \varepsilon} E_{\Omega_t}\left(u(\tilde{\gamma}_{t_n})^2 - u(\tilde{\gamma}_{t_{n-1}})^2 \,\Big|\, \mathcal{F}_{n-1}\right) + \sum_{k=0}^{n-2} \frac{S_{k+1}^2 - S_k^2}{F_k^* + \varepsilon},$$

$$E_{\Omega_t}\left(\frac{S_t^2}{F_t^* + \varepsilon}\,\Big|\,\mathcal{F}_{n-2}\right) \leq cE_{\Omega_t}\left(\frac{u(\tilde{\gamma}_{t_n})^2 - u(\tilde{\gamma}_{t_{n-1}})^2}{F_{n-1}^* + \varepsilon} + \frac{u(\tilde{\gamma}_{t_{n-1}})^2 - u(\tilde{\gamma}_{t_{n-2}})^2}{F_{n-2}^* + \varepsilon}\,\Bigg|\,\mathcal{F}_{n-2}\right) +$$

$$\sum_{k=0}^{n-3} \frac{S_{k+1}^2 - S_k^2}{F_k^* + \varepsilon},$$

$$\vdots$$

$$E_{\Omega_t}\left(\frac{S_t^2}{F_t^* + \varepsilon}\,\Big|\,\mathcal{F}_0\right) \leq cE_{\Omega_t}\left(\sum_{k=0}^{n-1} \frac{u(\tilde{\gamma}_{t_{k+1}})^2 - u(\tilde{\gamma}_{t_k})^2}{F_k^* + \varepsilon}\,\Bigg|\,\mathcal{F}_0\right)$$

$$= cE_{\Omega_t}\left(\sum_{k=0}^{n-1} \left|\frac{u(\tilde{\gamma}_{t_{k+1}})^2}{F_{k+1}^* + \varepsilon} - \frac{u(\tilde{\gamma}_{t_k})^2}{F_k^* + \varepsilon}\right|\right) + cE_{\Omega_t}\left(\sum_{k=0}^{n-1} u(\tilde{\gamma}_{t_{k+1}})^2\left(\frac{1}{F_k^* + \varepsilon} - \frac{1}{F_{k+1}^* + \varepsilon}\right)\right)$$

$$= cE_{\Omega_t}\left(\frac{u(\tilde{\gamma}_t)^2}{F_t^* + \varepsilon}\right) + cE_{\Omega_t}\left(\sum_{k=0}^{n-1} \frac{u(\tilde{\gamma}_{t_{k+1}})^2}{(F_{k+1}^* + \varepsilon)(F_k^* + \varepsilon)}(F_{k+1}^* - F_k^*)\right)$$

$$\leq cE_{\Omega_t}\left(\frac{(F^*)^2}{F_t^* + \varepsilon}\right) + cE_{\Omega_t}\left(\sum_{k=0}^{n-1} \frac{F_{k+1}^*}{F_k^* + \varepsilon}(F_{k+1}^* - F_k^*)\right).$$

When we let $t \to \infty$ and $\varepsilon \to 0$, the first term approaches $cE(F^*)$. In order to estimate the size of the second term, let

$$\Omega_\pi = \{\omega \in \Omega_t : F^*_{k+1}(\omega) \le F^*_k(\omega) + \varepsilon \text{ for } k = 0, 1, \ldots, n-1\}.$$

Then

$$\int_{\Omega_\pi} \sum_{k=0}^{n-1} \frac{F^*_{k+1}}{F^*_k + \varepsilon} (F^*_{k+1} - F^*_k) dP \le \int_{\Omega_\pi} \sum_{k=0}^{n-1} (F^*_{k+1} - F^*_k) dP$$

$$= \int_{\Omega_\pi} F^*_t dP \le E_{\Omega_t}(F^*) ,$$

and

$$\int_{\Omega \backslash \Omega_\pi} \sum_{k=0}^{n-1} \frac{F^*_{k+1}}{F^*_k + \varepsilon} (F^*_{k+1} - F^*_k) dP \le \int_{\Omega \backslash \Omega_\pi} \frac{F^*}{\varepsilon} \sum_{k=0}^{n-1} (F^*_{k+1} - F^*_k) dP$$

$$\le \frac{1}{\varepsilon} \int_{\Omega \backslash \Omega_\pi} (F^*)^2 dP ;$$

if we can show that the partition π can be chosen so that $P(\Omega \backslash \Omega_\pi)$ is arbitrarily small, then, since $F^* \in L^2(\Omega)$ and $\varepsilon > 0$ is fixed, the last term can be seen to be negligible and it will follow that

$$E_{\Omega_t}\left(\frac{S_t^2}{F^*_t + \varepsilon} \right) \le cE(F^*)$$

and hence

$$\| SF \|_{L^1(\Omega)} \le c \| F^* \|_{L^1(\Omega)} .$$

However, if for each $n = 1, 2, \ldots$ we let

$$A_n = \{\omega \in \Omega : F^*_{\frac{k+1}{2^n}t} > F^*_{\frac{k}{2^n}t} + \varepsilon \text{ for some } k = 0, 1, \ldots, 2^n - 1\},$$

then by the continuity of the sample paths

$$\Omega_t \subset \bigcup_{n=1}^{\infty} (\Omega_t \backslash A_n) ,$$

so that $P(A_n) \to 0$; therefore it is possible to find a partition π with the necessary property.

Theorem 7.3.2. 1. <u>If</u> $L \in (H^1)^*$, <u>then there is a function</u> $F \in H^2$ <u>such that</u>

(i) $L(f) = \int_\Gamma f(e^{i\theta})\overline{F(e^{i\theta})}dm(\theta)$ for all $f \in H^2$, and

(ii) $\operatorname{Re} F = \psi_1 + \tilde{\psi}_2$ for some $\psi_1, \psi_2 \in L^\infty_{\mathbf{R}}(\Gamma)$.

2. If $\phi = \psi_1 + \tilde{\psi}_2$ for some $\psi_1, \psi_2 \in L^\infty_{\mathbf{R}}(\Gamma)$, and $u = \mathscr{P}\phi$, then

$$\left|\int_\Gamma \phi dm\right| + \sup_{t\geq 0} \left\| E([u(\gamma_\tau) - u(\tilde{\gamma}_t)]^2 \middle| \tilde{\gamma}_t)\right\|_{L^\infty(\Omega)} < \infty$$

(so that ϕ has bounded mean oscillation and hence so does $\phi + i\tilde{\phi}$).

3. If $F \in H^2$ and

$$\sup_{t\geq 0} \left\| E(|F(\gamma_\tau) - F(\tilde{\gamma}_t)|^2 \middle| \tilde{\gamma}_t)\right\|_{L^\infty(\Omega)} < \infty,$$

then there is a constant c such that

$$\left|\int_\Gamma f(e^{i\theta})\overline{F(e^{i\theta})}dm(\theta)\right| \leq c\|f\|_{H^1} \quad \text{for all } f \in H^2.$$

(Since H^2 is dense in H^1, this action of F extends to a continuous linear functional on H^1.)

Moreover, the norms associated with these three statements are equivalent. Thus $(H^1)^*$ is isomorphic (but not isometrically isomorphic) to the space of all analytic functions on D whose boundary functions have bounded mean oscillation.

Proof. 1. Let us write $L = L_1 + iL_2$, where L_1 and L_2 are real-valued. When H^1 is regarded as a real Banach space, it may be identified with the subspace

$$X = \{(f, g) : g = \tilde{f}\}$$

of the real Banach space $L^1_{\mathbf{R}}(\Gamma) \times L^1_{\mathbf{R}}(\Gamma)$. Since L_1 is a real-valued continuous linear functional on X, it extends to an element $\hat{L}_1$ of the dual space of $L^1_{\mathbf{R}}(\Gamma) \times L^1_{\mathbf{R}}(\Gamma)$, and hence there are $\psi_1, \psi_2 \in L^\infty_{\mathbf{R}}(\Gamma)$ such that

$$\hat{L}_1(u, v) = \int_\Gamma u\psi_1 dm + \int_\Gamma v\psi_2 dm \quad \text{for all } u, v \in L^1_{\mathbf{R}}(\Gamma).$$

Recall that if $h \in L^2(\Gamma)$ has Fourier expansion

$$h(e^{i\theta}) = \sum_{n=-\infty}^{\infty} a_n e^{in\theta},$$

then its Fourier conjugate $\tilde{h}$ has expansion

$$\tilde{h}(e^{i\theta}) = \sum_{n=-\infty}^{\infty} \varepsilon_n a_n e^{in\theta},$$

where

$$\varepsilon_n = \begin{cases} 0 & \text{if } n = 0 \\ -i & \text{if } n > 0 \\ i & \text{if } n < 0 \end{cases}.$$

We see that $\int \tilde{h}\psi dm = -\int h\tilde{\psi} dm$ and $\tilde{\tilde{h}} = -h$ whenever $h, \psi \in L^2(\Gamma)$.

Thus if $u + i\tilde{u} \in H^2$,

$$L_1(u + i\tilde{u}) = \hat{L}_1(u, \tilde{u}) = \int_\Gamma u(\psi_1 - \tilde{\psi}_2)dm .$$

There is a standard way to recover a complex linear functional on a complex Banach space from its real part (which is a real linear functional on the associated real Banach space): we must have

$$L(f) = L_1(f) - iL_1(if) \quad \text{for all } f \in H^1.$$

Thus for $f = u + i\tilde{u} \in H^2$,

$$L(f) = \int_\Gamma u(\psi_1 - \tilde{\psi}_2)dm - i\int_\Gamma(-\tilde{u})(\psi_1 - \tilde{\psi}_2)dm = \int_\Gamma f(\psi_1 - \tilde{\psi}_2)dm.$$

Since also

$$-i\int f(\psi_1 - \tilde{\psi}_2)^{\sim} dm = -i\int f(\tilde{\psi}_1 + \psi_2)dm = -i\int (u + i\tilde{u})(\tilde{\psi}_1 + \psi_2)dm$$

$$= \int(\tilde{u} - iu)(\tilde{\psi}_1 + \psi_2)dm = \int\tilde{u}(\tilde{\psi}_1 + \psi_2)dm - i\int u(\tilde{\psi}_1 + \psi_2)dm$$

$$= \int u(\psi_1 - \tilde{\psi}_2)dm + i\int\tilde{u}(\psi_1 - \tilde{\psi}_2)dm = \int f(\psi_1 - \tilde{\psi}_2)dm ,$$

we have

$$\int f[(\psi_1 - \tilde{\psi}_2) + i(\psi_1 - \tilde{\psi}_2)^{\sim}]^{-} dm = 2L(f) .$$

We may take $\xi_1 = \frac{1}{2}\psi_1$, $\xi_2 = \frac{1}{2}\psi_2$, and $F = (\xi_1 + \tilde{\xi}_2) + i(\xi_1 + \tilde{\xi}_2)^{\sim}$ to complete the proof of statement (1).

2. Clearly each $\xi \in L^{\infty}_{R}(\Gamma)$ satisfies the stated condition. This part of the Theorem is therefore an easy consequence of Corollary 7.2.7 and Proposition 7.2.11.

3. Let us suppose that $F \in H^2$ satisfies the stated condition (so that its boundary function has bounded mean oscillation), and let $f \in H^2$ be given. By Lemma 7.2.12 and the polarization identity

$$\langle x, y\rangle = \frac{1}{4}[\|x+y\|^2 - \|x-y\|^2 + i\|x+iy\|^2 - i\|x-iy\|^2],$$

we have

$$\left|\int_{\Gamma} f(e^{i\theta})\overline{F(e^{i\theta})}dm(\theta)\right| = \frac{2}{\pi}\left|\iint_D f'(z)\overline{F'(z)}\log\left(\frac{1}{|z|}\right)dA(z)\right| .$$

As in the preceding section, we rewrite this as

$$\frac{2}{\pi}\left|\iint_D f'(z)\overline{F'(z)}\int_0^{\infty}\delta(0,t,z)dtdA(z)\right| = \frac{2}{\pi}\left|\int_0^{\infty}\iint_D f'(z)\overline{F'(z)}d\mu_t(z)dt\right|$$

$$= \frac{2}{\pi}\left|\int_0^{\infty}\int_{\{t<\tau\}} f'(\tilde{\gamma}_t)\overline{F'(\tilde{\gamma}_t)}dPdt\right| = \frac{2}{\pi}\left|\int_{\Omega}\int_0^{\tau(\omega)} f'(\tilde{\gamma}_t(\omega))\overline{F'(\tilde{\gamma}_t(\omega))}dtdP\right|.$$

We again use the <u>square functions</u>

$$Sf(\omega) = \left[\int_0^{\tau(\omega)}|f'(\tilde{\gamma}_s)|^2 ds\right]^{\frac{1}{2}} \text{ and } S_t f(\omega) = \left[\int_0^{t\wedge\tau(\omega)}|f'(\tilde{\gamma}_s)|^2 ds\right]^{\frac{1}{2}} .$$

Let $K_t = (S_t f)^2$ and $K = (Sf)^2$, let $\varepsilon > 0$, and use the Schwarz Inequality to find that

$$(*) \quad \left|\int_{\Gamma} f(e^{i\theta})\overline{F(e^{i\theta})}dm\right| = \frac{2}{\pi}\left|\int_{\Omega}\int_0^{\tau}\left[\frac{f'(\tilde{\gamma}_t)}{\sqrt[4]{(K_t+\varepsilon)}}\right]\sqrt[4]{(K_t+\varepsilon)}\,\overline{F'(\tilde{\gamma}_t)}\,dtdP\right|$$

$$\le \frac{2}{\pi}\left[\int_{\Omega}\int_0^{\tau}\frac{|f'(\tilde{\gamma}_t)|^2}{\sqrt{(K_t+\varepsilon)}}dtdP\right]^{\frac{1}{2}}\left[\int_{\Omega}\int_0^{\tau}\sqrt{(K_t+\varepsilon)}\,|F'(\tilde{\gamma}_t)|^2 dtdP\right]^{\frac{1}{2}};$$

denote the first of these factors by $\sqrt{I}$ and the second by $\sqrt{II}$.

Since $\tilde{\gamma}_t(\omega)$ is almost surely a continuous function of t, for $t < \tau$ and almost everywhere dP we have

$$\frac{d}{dt}(K_t + \varepsilon) = |f'(\tilde{\gamma}_t)|^2 ,$$

and hence

$$I = E\left(\int_0^{\tau} \frac{\frac{d}{dt}(K_t + \varepsilon)}{K_t + \varepsilon} dt \right) = E(2\sqrt{(K_t + \varepsilon)}).$$

Letting $\varepsilon \to 0$,

$$I = 2\|S_t f\|_{L^1(\Omega)} \leq 2\|Sf\|_{L^1(\Omega)} .$$

On the other hand, integration by parts shows that (after $\varepsilon \to 0$)

$$II = \int_\Omega \int_0^\tau S_t f |F'(\tilde{\gamma}_t)|^2 dt dP$$

(**)

$$= \int_\Omega [(S_t f)(S_t F)^2 \Big|_0^\tau - \int_0^\tau (S_t F)^2 \frac{d}{dt}(S_t f) dt] dP$$

$$= \int_\Omega [(Sf)(SF)^2 - \int_0^\tau (S_t F)^2 \frac{d}{dt}(S_t f) dt] dP$$

$$= \int_\Omega [(SF)^2 \int_0^\tau \frac{d}{dt}(S_t f) dt - \int_0^\tau (S_t F)^2 \frac{d}{dt}(S_t f) dt] dP$$

$$= \int_\Omega \int_0^\tau [(SF)^2 - (S_t F)^2] \frac{d}{dt}(S_t f) dt dP$$

$$= \int_0^\infty \int_{\{t<\tau\}} E((SF)^2 - (S_t F)^2 \,|\, \mathcal{B}_t) \frac{d}{dt}(S_t f) dP dt$$

$$= \int_0^\infty \int_{\{t<\tau\}} E((SF)^2 - (S_t F)^2 \,|\, \tilde{\gamma}_t) \frac{d}{dt}(S_t f) dP dt$$

$$= \int_\Omega \int_0^\tau E(\int_t^\tau |F'(\tilde{\gamma}_s)|^2 ds \,|\, \tilde{\gamma}_t) \frac{d}{dt}(S_t f) dt dP$$

$$\leq c \int_\Omega \int_0^\tau \frac{d}{dt}(S_t f) dt dP = c \int_\Omega Sf dP = c\|Sf\|_{L^1(\Omega)},$$

by Proposition 7.2.15. Therefore

$$\left| \int_\Gamma f(e^{i\theta}) \overline{F(e^{i\theta})} dm(\theta) \right| \leq c\|Sf\|_{L^1(\Omega)},$$

and an application of the Lemma completes the proof.

The similarities of this proof to the analytical one (as clearly presented in [51], for example) are obvious: Lemma 7.2.12 replaces the Littlewood-Paley identity, the martingale square functions replace the Lusin S-function, essentially the same trick (*) appears in both contexts, and an integration by parts (**) replaces an application of Green's Theorem. Thus the content and form of the proof remain much the same, although its material is of course vastly different.

7.3.3 **Summary.** Let $\phi \in L^2_R(\Gamma)$, $u = \mathcal{P}\phi$, and $F = u + i\tilde{u}$. Then the following statements are equivalent:

1. ϕ has bounded mean oscillation, in that

$$\sup_I \frac{1}{m(I)} \int_I \left|\phi - \frac{1}{m(I)} \int_I \phi dm\right| dm < \infty .$$

2. F has bounded mean oscillation.

3. $\sup_I \frac{1}{m(I)} \int_I \left|\phi - \frac{1}{m(I)} \int_I \phi dm\right|^p dm < \infty$ for some p with $1 \leq p < \infty$.

4. For each short interval I there is a number α_I such that

$$\sup_I \frac{1}{m(I)} \int_I (\phi - \alpha_I)^2 dm < \infty .$$

5. There are positive numbers b and N such that for each subinterval I of Γ and each $t \geq N$

$$m\{e^{i\theta} \in I : \left|\phi(e^{i\theta}) - \frac{1}{m(I)} \int_I \phi dm\right| > t\} \leq m(I)e^{-bt} .$$

6. $\sup_{r,\theta} \int_\Gamma [\phi(e^{i\sigma}) - \int_\Gamma \phi(e^{i\psi})\mathcal{P}(r, \theta-\psi)dm(\psi)]^2 \mathcal{P}(r, \theta-\sigma)dm(\sigma) < \infty.$

7. $\sup_{z \in D} [\mathcal{P}|\phi + i\tilde{\phi}|^2(z) - |\mathcal{P}(\phi + i\tilde{\phi})(z)|^2] < \infty .$

8. $\sup_{z \in D} \int_\Omega [\phi(\gamma_{z,\tau}) - u(z)]^2 dP < \infty .$

9. $$\sup_{t\geq 0} \left\| E([u(\gamma_\tau) - u(\tilde{\gamma}_t)]^2 \middle| \tilde{\mathcal{B}}_t) \right\|_{L^\infty(\Omega)} = \sup_{t\geq 0} \left\| E([u(\gamma_\tau) - u(\tilde{\gamma}_t)]^2 \middle| \tilde{\gamma}_t) \right\|_{L^\infty(\Omega)} = \sup_{t\geq 0} \left\| E(u(\gamma_\tau)^2 - u(\tilde{\gamma}_t)^2 \middle| \tilde{\gamma}_t) \right\|_{L^\infty(\Omega)} < \infty .$$

10. $\sup_{t\geq 0} \left\| E(|F(\gamma_\tau) - F(\tilde{\gamma}_t)|^2 \middle| \tilde{\gamma}_t) \right\|_{L^\infty(\Omega)} < \infty .$

11. $\sup_{z \in D} \iint_D |F'(w)|^2 g(z, w)dA(w) = \sup_{z \in D} \iint_D |\nabla u(w)|^2 g(z,w)dA(w) < \infty.$

12. $\sup_{t\geq 0} \left\| E(\int_t^\tau |F'(\tilde{\gamma}_s)|^2 ds \middle| \tilde{\gamma}_t) \right\|_{L^\infty(\Omega)} = \sup_{t\geq 0} \left\| E(\int_t^\tau |\nabla u(\tilde{\gamma}_s)|^2 ds \middle| \tilde{\gamma}_t) \right\|_{L^\infty(\Omega)} < \infty.$

13. There is a constant c such that

$$\left|\int_\Gamma f(e^{i\theta})\phi(e^{i\theta})dm(\theta)\right| \le c\|f\|_{H^1} \quad \text{for all } f \in H^2.$$

14. There is a constant c such that

$$\left|\int_\Gamma f(e^{i\theta})\overline{F(e^{i\theta})}dm(\theta)\right| \le \|f\|_{H^1} \quad \text{for all } f \in H^2.$$

15. There are $\psi_1, \psi_2 \in L^\infty_R(\Gamma)$ such that $\phi = \psi_1 + \tilde{\psi}_2$.

16. There is a constant c such that

$$\int_{1-h}^1 \int_{\theta_0}^{\theta_0+h} (1 - r^2)|F'(re^{i\theta})|^2 r d\theta dr \le ch$$

for all $h \in [0, 1)$ and all $\theta_0 \in [0, 2\pi)$.

References

1. Hirotada Anzai. Mixing up property of Brownian motion, Osaka Math. J. 2 (1950), 51-8.
2. Sheldon Axler. A real variable characterization of H^1, unpublished manuscript, 1975.
3. R. M. Blumenthal. An extended Markov property, Trans. Amer. Math. Soc. 85 (1957), 52-72.
4. D. L. Burkholder. Martingale transforms, Ann. Math. Stat. 37 (1966), 1494-1504.
5. ——— Harmonic analysis and probability, Studies in harmonic analysis, to appear.
6. D. L. Burkholder and R. F. Gundy. Extrapolation and interpolation of quasi-linear operators on martingales, Acta Math. 124 (1970), 249-304.
7. D. L. Burkholder, R. F. Gundy, and M. L. Silverstein. A maximal function characterization of the class H^p, Trans. Amer. Math. Soc. 157 (1971), 137-53.
8. L. Carleson. An interpolation problem for bounded analytic functions, Amer. J. Math. 80 (1958), 921-30.
9. ——— Interpolations by bounded analytic functions and the corona problem, Annals of Math. 76 (1962), 547-59.
10. Ronald R. Coifman. A real variable characterization of H^p, Studia Math. 51 (1974), 269-74.
11. Burgess Davis. On the integrability of the martingale square function, Israel J. Math. 8 (1970), 187-90.
12. ——— On the weak type (1, 1) inequality for conjugate functions, Proc. Amer. Math. Soc. 44 (1974), 307-11.
13. ——— Picard's theorem and Brownian motion, Trans. Amer. Math. Soc. 213 (1975), 353-62.

14. J. L. Doob. Stochastic Processes, John Wiley and Sons, Inc., New York, 1953.

15. ——— Semimartingales and subharmonic functions, Trans. Amer. Math. Soc. 77 (1954), 86-121.

16. Peter L. Duren. Theory of H^p spaces, Academic Press, New York, 1970.

17. A. Dvoretzky, P. Erdős, and S. Kakutani. Double points of paths of Brownian motion in n-space, Acta Sci. Math. Szeged 12 (1950), 75-81.

18. E. B. Dynkin. Markov processes, I and II, Academic Press Inc., New York, 1965.

19. C. Fefferman and E. M. Stein. H^p spaces of several variables, Acta Math. 129 (1972), 137-93.

20. David Freedman. Brownian motion and diffusion, Holden-Day, San Francisco, 1971.

21. Adriano M. Garsia. Martingale inequalities: Seminar notes on recent progress W. A. Benjamin, Inc., Reading, Mass., 1973.

22. George Gasper, Jr. On the Littlewood-Paley g-function and the Lusin S-function, Trans. Amer. Math. Soc. 134 (1968), 385-403.

23. R. K. Getoor and M. J. Sharpe. Conformal martingales, Inventiones Math. 16 (1972), 271-308.

24. G. H. Hardy and J. E. Littlewood. A maximal theorem with function-theoretic applications, Acta Math. 54 (1930), 81-116.

25. Eberhard Hopf. Ergodentheorie, J. Springer, Berlin, 1937.

26. G. A. Hunt. Some theorems concerning Brownian motion, Trans. Amer. Math. Soc. 81 (1956), 294-319.

27. Kiyoshi Itô and Henry P. McKean, Jr. Diffusion processes and their sample paths, Academic Press Inc., New York, 1965.

28. F. John and L. Nirenberg. On functions of bounded mean oscillation, Comm. Pure and Applied Math. 14 (1961), 415-26.

29. Shizuo Kakutani. Two-dimensional Brownian motion and harmonic functions, Proc. Imp. Acad. Tokyo 20 (1944), 706-14.

30. John G. Kemeny, J. Laurie Snell, and Anthony W. Knapp. Denumerable Markov chains, D. Van Nostrand Co., Inc., Princeton, N. J., 1966.

31. A. Khintchine. Über dyadische Brüche, Math. Zeit. 18 (1923), 109-16.

32. ——— Asymptotische Gesetze der Wahrscheinlichkeitsrechnung, Chelsea, New York, 1948.

33. A. N. Kolmogorov. Sur les fonctions harmoniques conjugées et les séries de Fourier, Fund. Math. 7 (1925), 24-9.

34. John Lamperti. Probability, W. A. Benjamin, Inc., New York, 1966.

35. Paul Lévy. Théorie de l'addition des variables aléatoires, Gauthier-Villars, Paris, 1937.

36. ——— Processus stochastiques et mouvement Brownien, Gauthier-Villars, Paris, 1948.

37. J. E. Littlewood and R. E. A. C. Paley. Theorems on Fourier series and power series, I, J. London Math. Soc. 6 (1931), 230-3.

38. ——— Theorems on Fourier series and power series, II, Proc. London Math. Soc. 42 (1936), 52-89.

39. ——— Theorems on Fourier series and power series, III, Proc. London Math. Soc. 43 (1937), 105-26.

40. Michel Loève. Probability theory, D. Van Nostrand Co., Inc., Princeton, N. J., 1963.

41. N. Lusin. Sur une propriété des fonctions à carré sommable, Bull. Calcutta Math. Soc. 20 (1930), 139-54.

42. J. Marcinkiewicz and A. Zygmund. A theorem of Lusin, Duke Math. J. 4 (1938), 473-85.

43. H. P. McKean, Jr. Stochastic integrals, Academic Press Inc., New York, 1969.

44. Paul A. Meyer. Probability and potentials, Blaisdell Publishing Co., Waltham, Mass., 1966.

45. Edward Nelson. Dynamical theories of Brownian motion, Princeton University Press, Princeton, N. J., 1967.

46. D. S. Ornstein and P. Shields. Mixing Markov shifts of kernel type are Bernoulli, Advances in Math. 10 (1973), 143-6.

47. R. E. A. C. Paley and Norbert Wiener. Fourier transforms in the complex domain, Amer. Math. Soc. Colloq. Pub. 19, Amer. Math. Soc., New York, 1934.

48. K. R. Parthasarathy. Probability measures on metric spaces, Academic Press Inc., New York, 1967.

49. M. Riesz. Über Potenzreihen mit vorgeschriebenen Anfangsgliedern, Acta Math. 42 (1920), 145-71.

50. Paul Arthur Schilpp, ed. Albert Einstein: Philosopher-scientist, I and II, Harper and Brothers, New York, 1959.

51. Joel Shapiro and Allen Shields. Three characterizations of BMO, unpublished seminar notes, Michigan State University, 1974.

52. D. C. Spencer. A function-theoretic identity, Amer. J. Math. 65 (1943), 147-60.

53. A. E. Taylor. Weak convergence in the spaces H^p, Duke Math. J. 17 (1950), 409-18.

54. A. E. Taylor. Banach spaces of functions analytic in the unit circle, I, Studia Math. 11 (1950), 145-70.

55. ——— Banach spaces of functions analytic in the unit circle, II, Studia Math. 12 (1951), 25-50.

56. G. E. Uhlenbeck and L. S. Ornstein. On the theory of Brownian motion, Phys. Rev. 36 (1930), 823-41.

57. Norbert Wiener. The homogeneous chaos, Amer. J. Math. 60 (1938), 897-936.

58. Norbert Wiener, Armand Siegel, Bayard Rankin, and William Ted Martin. Differential space, quantum systems, and prediction, M. I. T. Press, Cambridge, Mass., 1966.

59. Antoni Zygmund. Trigonometric series, I and II, Cambridge University Press, Cambridge, 1959.

Index